THE ECONOMIC CONSEQUENCES OF A CATASTROPHIC EARTHQUAKE

PROCEEDINGS OF A FORUM
AUGUST 1 AND 2, 1990

Committee on Earthquake Engineering

Division of Natural Hazard Mitigation

Commission on Engineering and Technical Systems

National Research Council

NATIONAL ACADEMY PRESS

Washington, D.C. 1992

NOTICE: The project that is the subject of this report was approved by the Governing Board of the National Research Council, whose members are drawn from the councils of the National Academy of Sciences, the National Academy of Engineering, and the Institute of Medicine. The members of the committee responsible for the report were chosen for their special competences and with regard for appropriate balance.

This report has been reviewed by a group other than the authors according to procedures approved by a Report Review Committee consisting of members of the National Academy of Sciences, the National Academy of Engineering, and the Institute of Medicine.

Any opinions, findings, and conclusions or recommendations expressed in this report are those of the committee and do not necessarily reflect the views of the sponsoring agencies.

Library of Congress Catalog Card Number 92-60712
International Standard Book Number 0-309-04639-4

Copies of this report are available from:
National Academy Press
2101 Constitution Avenue, N.W.
Washington, D.C. 20418

S-481

Printed in the United States of America

COMMITTEE ON EARTHQUAKE ENGINEERING

GEORGE W. HOUSNER (*Chair*), California Institute of Technology, Pasadena
KEIITI AKI, Department of Geological Sciences, University of Southern California, Los Angeles
CHRISTOPHER ARNOLD, Building Systems Development, Inc., San Mateo, California
JAMES E. BEAVERS, Martin Marietta Energy Systems, Inc., Oak Ridge, Tennessee
RAY W. CLOUGH, Department of Civil Engineering, University of California, Berkeley
C. B. CROUSE, Dames & Moore, Seattle, Washington
JOANNE NIGG, Disaster Research Center, University of Delaware, Newark
METE A. SOZEN, Department of Civil Engineering, University of Illinois, Urbana-Champaign
ROBERT V. WHITMAN, Department of Civil Engineering, Massachusetts Institute of Technology, Cambridge
JAMES K. WIGHT, Department of Civil Engineering, University of Michigan, Ann Arbor
T. LESLIE YOUD, Department of Civil Engineering, Brigham Young University, Provo, Utah

Liaison Representatives

WILLIAM H. ALLERTON, Division of Inspections, Federal Energy Regulatory Commission, Washington, D.C.
WILLIAM A. ANDERSON, Program Director, Division of Biological and Critical Systems, National Science Foundation, Washington, D.C.
C. CHESTER BIGELOW, Division of Advanced Technology Development, U.S. Department of Energy, Washington, D.C.
STEPHEN BOYCE, Air Force Office of Scientific Research, Bolling Air Force Base, Washington, D.C.
MANMOHAN S. CHAWLA, Naval Facilities Engineering Command, Washington, D.C.
DAE H. CHUNG, Nuclear Systems Safety Program, Lawrence Livermore National Laboratory, Livermore, California
FRED COLE, Office of U.S. Foreign Disaster Assistance, Agency for International Development, Washington, D.C.
JAMES COOPER, Federal Highway Administration, U.S. Department of Transportation, Washington, D.C.
JAMES F. COSTELLO, Mechanical/Structural Engineering Branch, Office of Nuclear Regulatory Research, U.S. Nuclear Regulatory Commission, Washington, D.C.
RICHARD F. DAVIDSON, Civil Engineering, Geotechnical Branch, U.S. Army Corps of Engineers, U.S. Department of the Army, Washington, D.C.

G. ROBERT FULLER, Compliance Branch, U.S. Department of Housing and Urban Development, Washington, D.C.
WALTER W. HAYS, Office of Earthquakes, Volcanoes, and Engineering, U.S. Geological Survey, Reston, Virginia
JAMES R. HILL, Natural Phenomena Hazards Mitigation Program, U.S. Department of Energy, Washington, D.C.
PAUL KRUMPE, Office of U.S. Foreign Disaster Assistance, Agency for International Development, Washington, D.C.
H. S. LEW, Center for Building Technology, National Institute of Standards and Technology, Gaithersburg, Maryland
DON LINGER, Strategic Structures Branch, Defense Nuclear Agency, Washington, D.C.
S. C. LIU, Earthquake Hazard Mitigation, Division of Biological and Critical Systems, National Science Foundation, Washington, D.C.
UGO MORELLI, Office of Natural and Technological Hazards, Federal Emergency Management Agency, Washington, D.C.
KENNETH J. SULLIVAN, Office of Natural and Technological Hazards Programs, State and Local Programs and Support, Federal Emergency Management Agency, Washington, D.C.
J. LAWRENCE VON THUN, Bureau of Reclamation, U.S. Department of the Interior, Denver, Colorado
SPENCER WU, Air Force Office of Scientific Research, U.S. Department of the Air Force, Washington, D.C.
EDWARD YOUNGER, Structural Engineering Service, U.S. Department of Veterans Affairs, Washington, D.C.
ARTHUR ZEIZEL, Office of Natural and Technological Hazards Programs, State and Local Programs and Support, Federal Emergency Management Agency, Washington, D.C.

Staff

RILEY M. CHUNG, Director
SUSAN R. MCCUTCHEN, Administrative Assistant
SHIRLEY J. WHITLEY, Project Assistant
EDWARD LIPP, Editor

Technical Editor

CAROLETTA LOWE, Editorial Concepts, Columbia, Maryland

ADVISORY GROUP FOR THE FORUM ON EARTHQUAKE ECONOMIC ISSUES

JOANNE NIGG (*Chair*), Disaster Research Center, University of Delaware, Newark
RICHARD N. BOISVERT, Department of Agricultural Economics, Cornell University, Ithaca, New York
RICHARD K. EISNER, Bay Area Regional Earthquake Preparedness Project, Oakland, California
HOWARD KUNREUTHER, Department of Decision Sciences, The Wharton School, University of Pennsylvania, Philadelphia
SHIRLEY MATTINGLY, Office of Emergency Management, City of Los Angeles, California
JEROME MILLIMAN (retired), Bureau of Economic and Business Research, University of Florida, Gainesville
RISA PALM, Graduate School, University of Colorado, Boulder
WILLIAM J. PETAK, Institute of Safety and Systems Management, University of Southern California, Los Angeles

Liaison Representatives

GARY JOHNSON, Office of Natural and Technological Hazards Programs, Federal Emergency Management Agency, Washington, D.C.
EUGENE L. LECOMTE, The Earthquake Project, National Committee on Property Insurance, Boston, Massachusetts
JAMES TAYLOR, Insurance Support Services, Federal Insurance Administration, Federal Emergency Management Agency, Washington, D.C.

SPEAKERS

CHRISTOPHER ARNOLD, Building Systems Development, Inc., San Mateo, California
ROBERT G. CHAPPELL, State and Local Programs and Support, Federal Emergency Management Agency, Washington, D.C.
LEONARD K. CHENG, Department of Economics, University of Florida, Gainesville
HAROLD COCHRANE, Department of Economics, Colorado State University, Fort Collins
NEIL DOHERTY, Department of Decision Sciences, The Wharton School, University of Pennsylvania, Philadelphia
RONALD EGUCHI, Dames and Moore, Los Angeles, California
DON G. FRIEDMAN, Corporate Strategy and Research, Travelers Insurance Company, Hartford, Connecticut
ROBERT M. HAMILTON, U.S. Geological Survey, Reston, Virginia

ROBERT W. KLING, Department of Economics, Colorado State University, Fort Collins
HOWARD KUNREUTHER, Department of Decision Sciences, The Wharton School, University of Pennsylvania, Philadelphia
TAPAN MUNROE, Pacific Gas and Electric Company, San Francisco, California
RICHARD J. ROTH, Jr., California Department of Insurance, Los Angeles
BARBARA D. STEWART, Stewart Economics, Inc., Chapel Hill, North Carolina
KATHLEEN TIERNEY, Disaster Research Center, University of Delaware, Newark
L. THOMAS TOBIN, Seismic Safety Commission, Sacramento, California
ANTHONY M. YEZER, Department of Economics, George Washington University, Washington, D.C.

Original transcription of Proceedings by C.A.S.E.T. Associates, Fairfax, Virginia

Table of Contents

List of Illustrations

Figures:

Tables:

The National Academy of Sciences is a private, nonprofit, self-perpetuating society of distinguished scholars engaged in scientific and engineering research, dedicated to the furtherance of science and technology and to their use for the general welfare. Upon the authority of the charter granted to it by the Congress in 1863, the Academy has a mandate that requires it to advise the federal government on scientific and technical matters. Dr. Frank Press is president of the National Academy of Sciences.

The National Academy of Engineering was established in 1964, under the charter of the National Academy of Sciences, as a parallel organization of outstanding engineers. It is autonomous in its administration and in the selection of its members, sharing with the National Academy of Sciences the responsibility for advising the federal government. The National Academy of Engineering also sponsors engineering programs aimed at meeting national needs, encourages education and research, and recognizes the superior achievements of engineers. Dr. Robert M. White is president of the National Academy of Engineering.

The Institute of Medicine was established in 1970 by the National Academy of Sciences to secure the services of eminent members of appropriate professions in the examination of policy matters pertaining to the health of the public. The Institute acts under the responsibility given to the National Academy of Sciences by its congressional charter to be an adviser to the federal government and, upon its own initiative, to identify issues of medical care, research, and education. Dr. Kenneth Shine is president of the Institute of Medicine.

The National Research Council was established by the National Academy of Sciences in 1916 to associate the broad community of science and technology with the Academy's purposes of furthering knowledge and of advising the federal government. Functioning in accordance with general policies determined by the Academy, the Council has become the principal operating agency of both the National Academy of Sciences and the National Academy of Engineering in providing services to the government, the public, and the scientific and engineering communities. The Council is administered jointly by both Academies and the Institute of Medicine. Dr. Frank Press and Dr. Robert M. White are chairman and vice-chairman, respectively, of the National Research Council.

The Economic Consequences of a Catastrophic Earthquake

Executive Summary

BACKGROUND

Since Congress enacted the National Earthquake Hazard Reduction Act in 1977, substantial progress has been made in the area of earthquake hazard identification and strategies to deal with earthquake-related problems. The national program, coordinated by the Federal Emergency Management Agency (FEMA), with participation by the National Science Foundation (NSF), the U.S. Geological Survey (USGS), and the National Institute of Standards and Technology, addresses various issues related to earthquake hazards (e.g., risk assessment, prediction or forecasting, mitigation measures such as improvements of building codes, disaster response, preparedness planning, and recovery).

While many earthquake hazard-reduction issues have been addressed by the national program during the past 14 years, one issue that warrants more attention is the *economic consequences of a catastrophic earthquake.* Efforts to manage earthquake hazards must include an assessment of the public and private sectors' ability to reduce and recover from earthquake-induced losses. Stricken communities and states should have the ability to maintain sufficient financial stability, thus allowing them to rebuild their economic bases following a catastrophic event.

It is important to recognize that no truly catastrophic earthquake--that is, one that affects production facilities, economic markets, and distribution systems in any significant manner--has occurred in a major population center in the United States since the 1906 San Francisco earthquake. A national concern about potential economic consequences, however, has been heightened in recent years as scientifically based probabilities of future earthquakes in urbanized areas have risen, as their time windows have shortened, and as the social and economic costs of recent earthquakes have increased dramatically.

To address this concern, the National Research Council requested the Committee on Earthquake Engineering (CEE) of the Division of Natural Hazard Mitigation to organize a forum on issues related to potential economic consequences--locally, regionally, and nationally--that could occur following a catastrophic earthquake in the United States. Support for this forum was provided by the Federal Insurance Administration of the FEMA and by the National Committee on Property Insurance.

An Advisory Group for the Forum on Earthquake Economic Issues, chaired by CEE member Dr. Joanne Nigg, was established to identify key issues to be addressed during the forum. The forum, held on August 1-2, 1990, in Washington, D.C., consisted of 15 invited presentations that reflect the state of the art of economic theory, economic modeling, hazard characterization, and societal impacts as they can be applied to concerns about earthquake consequences. The first day and a half of the forum consisted of these presentations and short discussion periods to clarify points made in the presentations. The last half day of the forum was devoted to a general discussion, in which both the presenters and the audience participated, of issues raised during the meeting.

The following topics were addressed:

1. What is known about the seismic risk nationally?
2. What are the likely categories of loss and damage in the event of a catastrophic earthquake?
3. What is the current state of the art of economic research on earthquake consequences?
4. To what extent does the location of the catastrophic event affect the types and extent of impacts?
5. To what extent can regional and interregional shifts in resources take place following a catastrophic earthquake?
6. How likely is a "ripple effect" to occur following a catastrophic earthquake, and what consequences may result from such a phenomenon?
7. How do current state and federal postearthquake policies and programs affect economic recovery capabilities of the public and private sectors?

Regardless of their specific topic areas, presenters were asked to keep the following three orienting questions in mind. These questions were also used to focus the general discussion on the second day of the forum.

1. In the event of a catastrophic earthquake in the United States, what are the potential economic losses within the impact area?
2. Under what conditions might local or regional earthquake losses result in *national* economic disruption?
3. What conditions, programs, or policies are capable of either exacerbating or reducing the amount of both local and national disruption that could take place?

The invited presentations, including the clarifying questions asked by audience members, are in the second major section of this report. The remainder of this executive summary summarizes the general discussion on the second day of the forum with respect to (1) the major themes identified and the range of views expressed on these themes and (2) the types of research that are needed to resolve problems associated with the thematic issues.

The forum was not intended to provide recommendations on policy matters or needed program initiatives. Its primary objective was to review the major issues related to economic impacts from large-scale earthquakes, to assess current capabilities--in terms of data bases and economic models--to predict these impacts, and to review current state and federal programs related to recovery from such events.

THEMATIC ISSUES

Five thematic issues were extensively discussed during the second day of the forum: (1) the capabilities of and requirements for economic modeling, (2) the nature and scope of possible social and economic impacts, (3) the likelihood of a national "ripple effect," (4) the relationship between insurance and mitigation, and (5) the need for a new approach to the problem.

Economic Modeling

Panelists and audience members identified the need to project various types of economic impacts from a catastrophic earthquake, including: direct economic losses due to destroyed or severely damaged buildings and other structures (such as dams and lifeline systems), direct economic losses due to damaged or destroyed contents of buildings and other private property, indirect economic losses due to disruption of the private sector (that is, business interruption), loss of revenues and increases in expenses for the public sector, and losses of individual and household income due to injury, death, or job interruption.

Direct Structural Loss Estimates

Reasonably sound loss-estimation methods currently exist to project direct damage to buildings from ground motion. However, two major limitations of these methodologies were identified. First, the models have been primarily developed for application to engineered buildings and are less reliable when being applied to nonengineered buildings or to other types of structures (for example, to lifeline systems). Estimates of loss, therefore, are better for large, multistory, commercial and residential buildings than for smaller or, often, nonengineered (one- or two-story) buildings or residential dwellings.

Second, currently available data bases required by these models to develop reliable loss estimates at the community level are inadequate on a national basis. This is especially true outside of California. The primary type of information that is unavailable is an *inventory of existing buildings and structures*. Such an inventory might include: ages of the structures, construction types (e.g., configuration, height, and materials), usage patterns, and number of inhabitants. From this structural inventory, response to ground-motion estimates, and hazard characterizations, projected damage estimates *could be made*, including economic costs for replacement and repair of the structures, following earthquakes of various magnitudes for specific communities. Without good inventory data, however, these damage estimates are, at best, educated guesses.

Direct Nonstructural Losses

The forum discussions illustrated that currently no one methodology is generally accepted for projecting likely damage to the contents of structures or to other types of private property that are likely to be affected by structural damage. The magnitude of the losses that could be associated with damage to production machinery, office equipment (including computers), inventories, or raw materials due to collapsed or partially collapsed structures is unknown. The indirect consequences of these types of losses (in terms of business disruption) are discussed below. However, damage or destruction of these resources has an immediate, direct economic impact. That is, equipment and supplies must be replaced before production can resume.

Two kinds of problems were raised concerning damage projections. First, models have not been developed that relate direct structural damage to nonstructural, content losses. Second, such models would have to rely on inventories that were even more extensive than those required to project structural losses. Although some modeling work has been undertaken to project nonstructural (or content) losses to a specific building, the work does not extend to classes of structures.

Economic Losses Due to Business Interruption

Business interruption can result from damage to the structure in which the business is located; damage to production or manufacturing equipment, office equipment, and inventory; loss of production materials due to losses experienced by a supplier whose facilities were also damaged; loss of electrical power or other lifeline services necessary to operate the facility; interruption of the transportation system to deliver supplies or finished products; loss of customers due to damage to their facilities or inability to access the facility; or loss of employees due to death or injury.

Little is empirically known about disaster-generated indirect economic effects on business in general, and less on the consequences for specific economic sectors. However, recent postearthquake research has indicated that few small businesses affected by an earthquake are likely to have business interruption insurance. Several participants stressed that this situation could result in an increase in the business failure rate following a large-magnitude earthquake.

The participants reported that some economic research has been undertaken on market interdependencies at a national level, but little effort appears to have been expended on this problem at a subnational level. These participants also indicated that to begin to sufficiently address this concern at either a regional or interregional level, data bases must be developed on the geographic distribution of trade flows to make the available models operational.

While many participants felt that such economic loss projections would be useful, there was little agreement on how accurate such estimates had to be to provide policy advice to decision makers in the public or private sectors.

Public Sector Economic Costs

A catastrophic earthquake could affect government at all levels, but especially at the local and state levels, by reducing future revenues, increasing current costs resulting from response activities, and increasing future costs resulting from recovery and reconstruction activities. However, there could also be revenue increases to offset these decreases: for example, construction permits and new business licenses (for construction-related businesses), and temporary occupancy fees (bed taxes) paid by temporary repair and relief personnel. Nevertheless, the extent to which these increases and decreases would be balanced, and for which levels of government, is yet to be explored.

The participants reported that no systematic research has been conducted on the overall economic effects of a major disaster on the public sector, much less on trying to project these impacts for a future catastrophic earthquake; nor do models or data bases currently exist to estimate these different types of economic costs to government. Also, despite the fact that emergency service professionals have expended a great deal of effort on emergency-response planning for a destructive earthquake during the past two decades, these plans have never been used to estimate governmental budgetary needs. These increased postearthquake governmental costs, as described at the forum, include: debris removal and disposal, urban search and rescue efforts, fire-fighting and hazardous-materials-event response, provision of emergency medical services, provision of temporary shelter, overtime for salaried governmental workers to perform a variety of operational and administrative services, and inability to invest in new, productive projects because of repair and reconstruction costs associated with publicly owned damaged facilities.

Similarly, the participants noted that no methodology exists for estimating the impact that a large-scale disaster could have on government revenues. Revenue losses could result from changes in property tax assessments (due to structural damage), reduction in both business and personal income taxes, declines in purchases resulting in lower sales taxes, declines in bed taxes due to a decline in tourism, and fewer user fees, among others.

Another unknown factor, as reported at the forum, is the extent to which the ability of a community to issue municipal bonds to fund government-sponsored development projects would be affected by a large earthquake. Some participants stressed the importance of this issue, since communities may have to pass additional bond measures to fund reconstruction projects for government-owned buildings and facilities. If currently funded projects are significantly damaged, resulting in the need to expend additional public monies to complete the project, postearthquake reconstruction projects may be further delayed because of inability to raise capital (e.g., through the sale of local bonds).

Personal and Household Economic Losses

Some presenters indicated that, following a federally declared disaster, households are generally able to recover; but recovery is affected by the nature of the losses, the degree of assistance provided, and the preearthquake socioeconomic status of the household and the community.

Insurance industry representatives at the forum expressed concern about the consequences to the industry if the need arises to cover a wide variety of personal and household loss claims from a catastrophic earthquake. In addition to shake damage, which is covered by an earthquake insurance policy, there may be additional indirect losses, such as workers' compensation, medical costs, life insurance, liability coverage, and automobile damage. Such a catastrophic loss could erode insurers' surpluses, potentially resulting in an availability crisis (i.e., a market failure). The participants reported that while some information is available to estimate the extent of "covered" losses (i.e., covered by the insurance industry), no information currently exists to project *all* losses (including those that are not covered by insurance) for these categories.

The Nature and Scope of Earthquake Impacts

The participants noted that almost no solid empirical data are currently available regarding the nature, scope, and duration of likely economic and social impacts that would result from a catastrophic earthquake. They further identified a definite need to better understand these effects on several dimensions: direct and indirect losses and costs, economic and social losses and costs, impacts for various governmental levels (local, substate, state, regional, and national), and the consequences of disruption and dislocation for households, businesses, and communities.

In addition to the specific modeling issues addressed above, many participants expressed the belief that net economic impacts of catastrophic events are likely to be negative, at least for a short time at the substate level; but the likely duration and distribution of these impacts, especially for various economic sectors (e.g., manufacturing and retailing) and for specific social groups is largely unknown.

One major issue surfaced--the need to address the *redistribution effects and consequences* of large earthquakes. Even if impacts are not felt beyond the local or substate area, models need to be developed and attention paid to the "winners" and "losers" of these redistribution effects between, as well as within, economic or business sectors. Similarly, ethnic and class groups will be differentially affected by a catastrophic earthquake, resulting in their differential ability to recover. While aggregate changes in the economic health of an area may be only temporary, sectoral changes may have longer-term consequences for the area's labor force, minority-owned businesses, or neighborhood property values. Such changes, as reported by some participants, could substantially affect the quality of life and the economic viability of the community. Mitigation and recovery planning must take these qualitative consequences, as well as the economic ones, into account.

The "Ripple Effect"

Extensive discussion focused on the likelihood that a catastrophic earthquake could generate widespread, unacceptable, negative economic consequences for the insurance industry and, by extension, for both the public and private sectors.

In part, the differences of opinion expressed during this discussion were philosophic. Those who maintained that economic consequences were more likely to be minor, local, and time-bound referred to the resilience of the market even during the worst economic periods in recent U.S. history (e.g., the stock market "crash" of October 1987). Those who painted a "doomsday" scenario argued that financial markets are extremely sensitive to any uncertain situation. For either argument, the participants agreed that with respect to large-scale natural disasters in recent history, there is very little empirical evidence available, other than data on immediate, short-term impacts, upon which to test either of these points of view. Some participants indicated that they do not believe a catastrophic earthquake would have a crippling effect on the national economy of the United States, and many stated that they felt the scope and duration of economic consequences could vary, depending on where such an earthquake occurs. Outside California (for example, in the central United States), greater direct structural and lifeline system damage would be expected (since a larger proportion of the existing building stock and lifeline-system was constructed without regard for seismic design), resulting in proportionally greater economic losses and social consequences.

Insurance and Mitigation

Because of the insurance industry's stated concerns about the erosion of their surplus capital as a consequence of a catastrophic earthquake, some of the forum discussion focused on the feasibility of earthquake insurance as a mitigation tool. Many participants believe that earthquake insurance, as it is currently offered, is *not* an efficient mechanism to reduce economic impacts on homeowners and businesses, for several reasons. In the highest-risk areas in California, for example, premiums are very high and deductibles very large, which has discouraged greater earthquake-insurance purchases. In high-risk areas outside of California, premiums may be substantially lower, but there is also a lower demand for earthquake insurance by the public, either because people do not perceive the risk to be sufficiently large to require coverage or because they are under the mistaken impression that their general homeowner's policy will cover earthquake-caused damage. The participants are in general agreement that the cost of inspecting residential structures to determine vulnerability or to verify levels of mitigation is prohibitive today, relative to the amount of premiums charged. Current insurance practices, therefore, do not tie rates to mitigation.

Much discussion focused on the need to develop a federal earthquake reinsurance fund to increase the insurance industry's ability to avoid a market failure in the event that a catastrophic earthquake occurs in a major metropolitan area. There was disagreement on the need for such a program, because it is unclear whether the insurance industry needs federal assistance to cover earthquake-related losses and because there is general skepticism about a national ripple effect. Nonetheless, participants on both sides of the issue agreed that if such a federal program were to be developed, the rate structure should be risk-based. That is, it should reflect the actual risk to the structure. Both sides also agreed that the rate structure should reward, through lower premiums, individual and community mitigation efforts to reduce that risk.

A New Approach

Throughout the forum discussions, it was apparent that the problem of anticipating the economic impacts and consequences from a catastrophic earthquake is not a problem that should be addressed either in a piecemeal fashion or by a single discipline or industrial sector. The participants nevertheless agreed that, unfortunately, much of the work done to date has suffered from one or both of these deficiencies. (One notable recent exception, however, is the FEMA-sponsored project, Loss-Reduction Provisions of a Federal Earthquake-Insurance Program.)

The participants further agreed that future research efforts should be multidisciplinary, involving university as well as public and private sector research teams. Because there was such a great emphasis among forum participants on the utility of these projects for government mitigation programs and fiscal planning, these research projects, as indicated by the participants, should also have multidisciplinary public- and private-sector advisory panels to assist in the refinement of research objectives and to provide ongoing input to the projects, including the provision of data for use in the economic models.

POTENTIAL RESEARCH TOPICS

Four types of research needs were identified by the forum participants: loss-estimation models, economic models, redistributional effects, and policy issues.

Loss-Estimation Models

1. Loss-estimation models would be improved if they could project direct losses to structures other than buildings, contents of buildings, possible fires following earthquakes, and the factors used in econometric models to describe economic change.

2. Applicability of the models will require inventories of structures and their contents. A methodology for developing such inventories could be formulated to make these models work.

3. Loss-estimation models will also require more-detailed characterization of the hazard, including projections of ground motion, from earthquakes of various magnitudes in high-risk areas.

Economic Models

4. Economic models and data are needed that can accurately project how a catastrophic earthquake would disrupt the flow of trade and commerce at regional, state, and substate levels.

5. These models should be developed based on the potential impacts of the critical "engines" that drive the economies in the various earthquake-risk regions of the country.

Redistributional Effects

6. Methodologies are needed that can project likely economic redistributional effects both between and within states.

7. Further, these methodologies should be refined to examine the differential redistributional effects across social groups and communities in possible impact areas to identify causes and, if necessary, solutions to negative impacts.

Policy Issues

8. A better understanding of the social and economic consequences resulting from different mixes of mitigation and hazard-reduction policies would allow local and state governments to evaluate issues such as what are

the benefits? who benefits and who loses? what are the costs? who pays these costs? what are the social effects of implementation? and how much public finance is justifiable?

9. An analysis is needed of the existing state and federal aid policies to assess their adequacy and consequences, especially for local governments.

10. Economic models that would provide government decision makers useful information for formulating mitigation and economic recovery policies can be formulated. These decision makers should be closely consulted in developing the needed level of precision of these models for use in their respective communities or regions.

1

What is Known About Seismic Risk Nationally?

Over 41 states in the United States are subject to damages and losses from earthquakes. Although California is the area most frequently associated with earthquake events, other areas of the country are also at risk, notably: the Puget Sound area in Oregon and Washington; the Wasatch fault area in Utah; the New Madrid fault area in the central United States; Charleston, South Carolina; the Boston-New York area; Alaska; Hawaii; and Puerto Rico. This chapter establishes the scientific basis for earthquake risks in these areas as well as "probabilities" (where such evidence exists) for when future earthquakes are likely to occur in various metropolitan centers.

Dr. Robert M. Hamilton is with the USGS. Dr. Hamilton's doctoral degree is in geophysics from the University of California at Berkeley. He has been with the U.S. Geological Survey since 1968 in a variety of roles, including Chief Geologist, Chief of the Office of Earthquake Studies, and coordinator of the Deep Continental Studies program. He is also past president of the Seismological Society of America. Beginning in August 1990, Dr. Hamilton accepted a 2-year post at the Secretariat of the International Decade for Natural Disaster Reduction (IDNDR) in Geneva.

PRESENTATION OF ROBERT M. HAMILTON

This presentation gives a brief overview of the nature of earthquake hazards in the United States. It will necessarily be a once-over-lightly treatment, but hopefully it will at least help to set the stage for some of the later chapters. Progress in understanding the nature of earthquake hazards in the United States has been very rapid over the last 25 years. Twenty-six years ago the concept of plate tectonics was only emerging. That concept now provides an understanding of the forces that cause earthquakes, and it is the basis for long-term earthquake predictions that have been made in some parts of the world, particularly around the rim of the Pacific Ocean, and especially in California.

As recently as 15 years ago, there was no geologic explanation for the four magnitude 8 earthquakes that struck New Madrid, Missouri, in 1811-1812 and devastated the area. Because they occurred in the middle of the North American plate, ordinary concepts of plate interaction did not fit. Geologists now understand that those earthquakes were caused by flexing within the North American plate, causing reactivation of a rift in the crust of the earth that formed over 500 million years ago. That rift has been reactivated repeatedly.

Just as there are differences between the causes of earthquakes in the western United States and in the East, there are also major differences in the way they affect structures and the way seismic-wave energy is propagated. For a given-magnitude earthquake, eastern earthquakes shake a much greater area than those in the West, about 20 times the area as a rough approximation. Although there are fewer earthquakes for a given period of time (that is, a lower frequency of earthquakes) in the East, the greater area that they affect somewhat offsets this lower frequency.

Figures 1-1 through 1-24 illustrate the ideas that give rise to understanding the cause of earthquakes. Also discussed are some of the effects of earthquakes that are relevant to this paper.

Figure 1-1 is a map of the world with dots that show the locations of earthquakes over a long period of time. They are not randomly distributed, but are located in narrow bands that define the boundaries of the large plates that make up the outer shell of the earth. The earthquakes are caused by colliding and scraping between these plates. The area of greatest earthquake activity is around the rim of the Pacific Ocean. In California, the San Andreas fault system makes up part of the boundary around this rim.

Figure 1-2 shows the plate boundary. As illustrated, the western margin of the United States is part of the boundary of the Pacific plate. The Pacific plate is drifting toward the northwest relative to California and the North American plate. Offshore California, including the Los Angeles area, is headed toward the Aleutian Islands. In several tens of millions of years, Los Angeles will become a suburb of San Francisco and then drift to be subducted into the interior of the earth, melting down and becoming a part of the Aleutian Island chain.

The central part of the United States, including the New Madrid area and the eastern United States, is in the middle of the North American plate. One problem in trying to understand central and eastern U.S. seismicity lies in understanding the origin of the forces that cause those earthquakes in the context of our general model of plate tectonics.

Figure 1-3 shows the processes of plate tectonics. Along the midocean ridges (e.g., along the middle of the Atlantic Ocean, or along the eastern part of the Pacific Ocean) partially molten material rises from the interior of the earth into the crust, where it cools to form new crust and then moves out in both directions. Where the plates collide, one plate goes under and one overrides. For example, along the Aleutian Arc, or along Japan, and anywhere in the western Pacific, the oceanic plate goes under the continental plate. Along those boundaries there are volcanoes, mountains, and a lot of earthquake activity. In some places, plates more or less scrape horizontally past each other, which is the case along the San Andreas fault.

Of course, the stresses that are associated with earthquakes are elastic forces. The stresses build up, much as when a yardstick is bent; when stresses become too great, the stick snaps. Similarly, elastic energy is released by the sudden movement of the plates. This energy propagates outward, shaking the ground and causing the earthquake effects that we are concerned about.

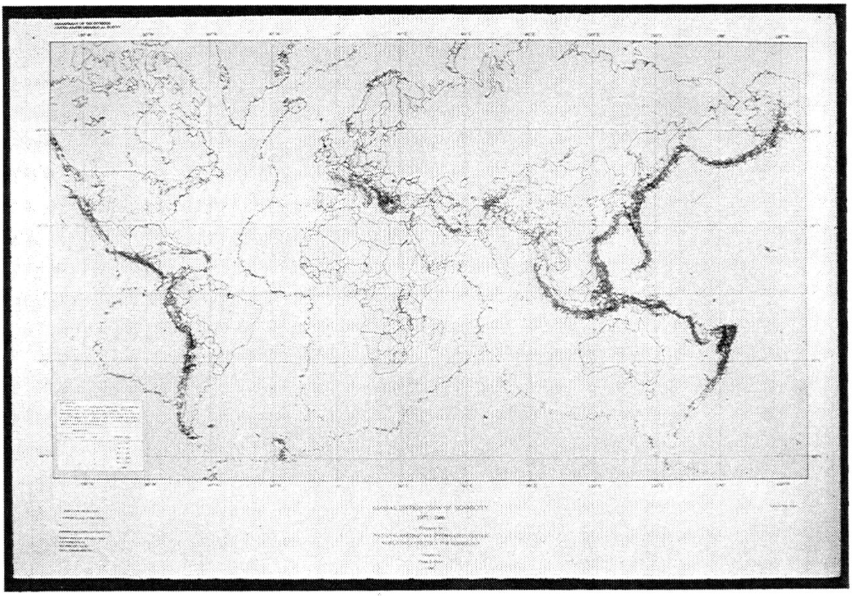

FIGURE 1-1 Global distribution of seismicity.

As previously noted, U.S. earthquake activity is associated with the North American plate (Figure 1-4). The plate boundary lies along the western margin of the continent; a lot of the plate-related activity extends into the western part of the United States as far as the Wasatch front in Utah. The Pacific plate is moving at about 8 cm per year relative to the North American plate, which is moving at about 2.5 cm per year. The New Madrid earthquake zone and all the other activity along the eastern margin is located within the North American plate.

Focusing on California, Figure 1-5 shows that earthquake activity is not in a narrow zone along the San Andreas fault, but rather occupies a fairly broad zone. There is an earthquake band, actually the boundary of the plate, that extends through the Gulf of California (Sea of Cortez) into the Imperial

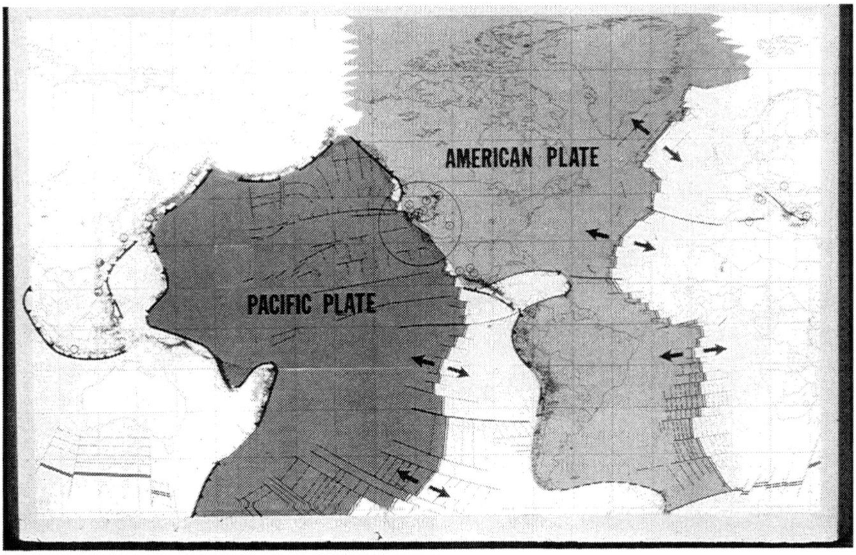

FIGURE 1-2 Earthquakes and global tectonics.

Valley, and along the San Andreas fault system (which, in reality, is a whole series of faults). Activity also runs through the Owens Valley into western Nevada, and there is a lot of activity along Cape Mendocino. It is a fairly complicated band of broad activity.

In southern California, particularly, things are very complicated. Figure 1-6 illustrates the concept that the San Andreas fault system is not just one boundary between the North American plate and the Pacific plate, but rather is a complex zone of blocks. In general, Los Angeles is moving north, but there are other blocks that are moving in different directions. In the transverse ranges (e.g., the Santa Barbara area) these blocks have to move out of the way to the west, which is why the mountains are there. The Mojave Desert block has to skip out to the east.

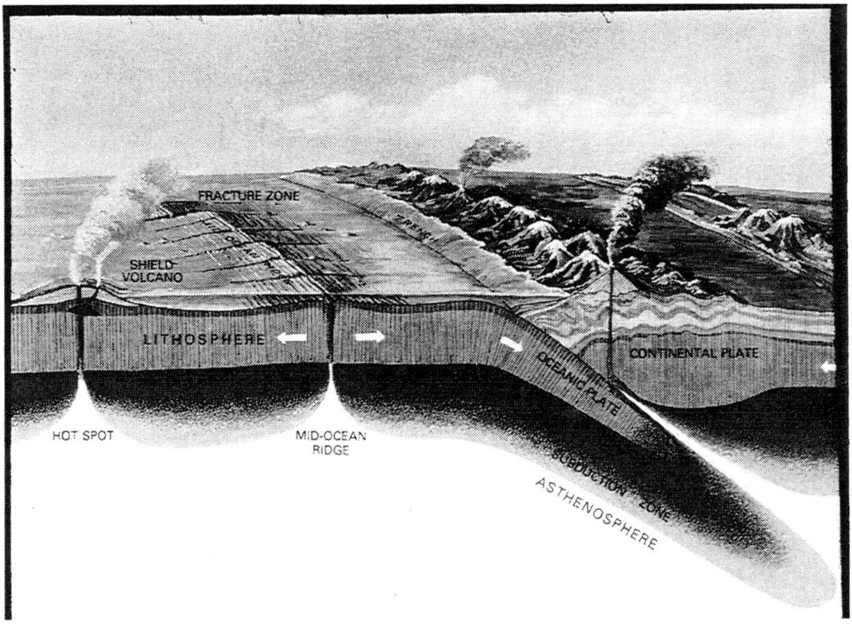

FIGURE 1-3 Schematic cross section illustrating plate tectonics processes.

The concept portrayed here is that the fault zone itself is a complex zone of faults and blocks that jostle each other. It is much like blocks of ice drifting on a river, where the underlying current of the river is moving the blocks along. As the blocks move, they shift to accommodate the presence of other blocks. That is a pretty good model to portray the crust of the earth: a bunch of blocks that are floating on a fluid interior and jostling around in order to accommodate that movement.

In California, there have been a series of very strong earthquakes. The 1906 earthquake broke the San Andreas from just north of San Juan Bautista toward Cape Mendocino. That was the last magnitude 8 earthquake in California. In 1857 another magnitude 8 earthquake that all the way from the Fort Tejon area south to the Mojave Desert. There was a magnitude near-8 earthquake in 1872 in the Owens Valley.

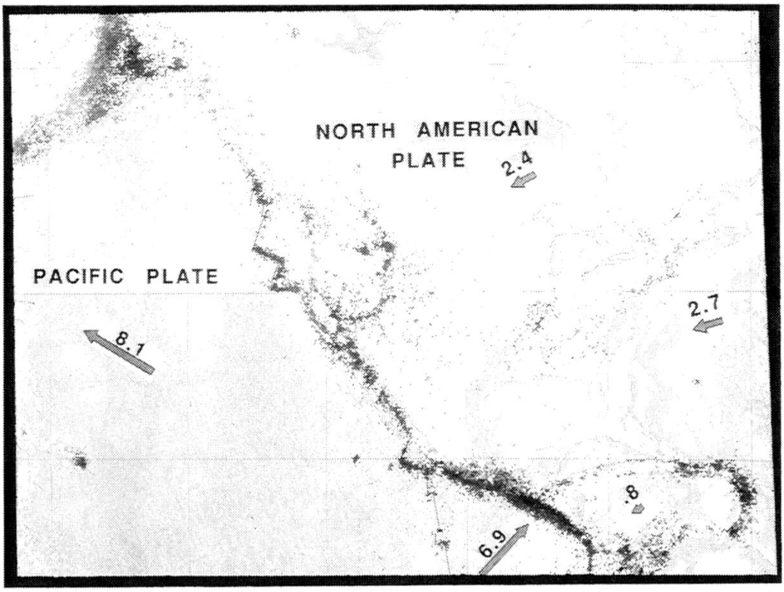

FIGURE 1-4 The North American plate.

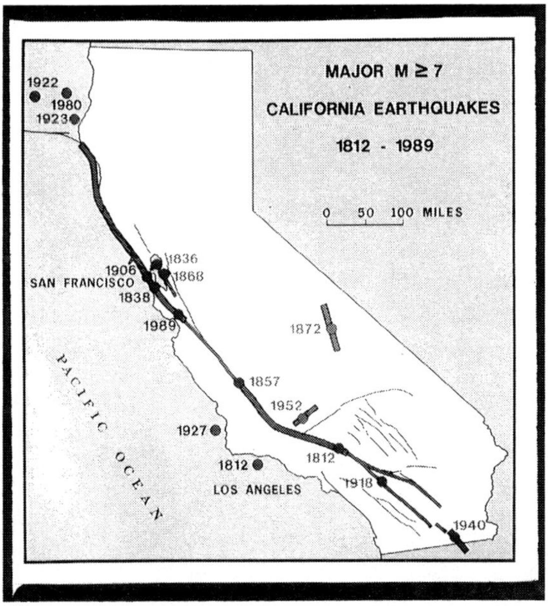

FIGURE 1-5 Major (M ≥ 7) California Earthquakes (1812-1989).

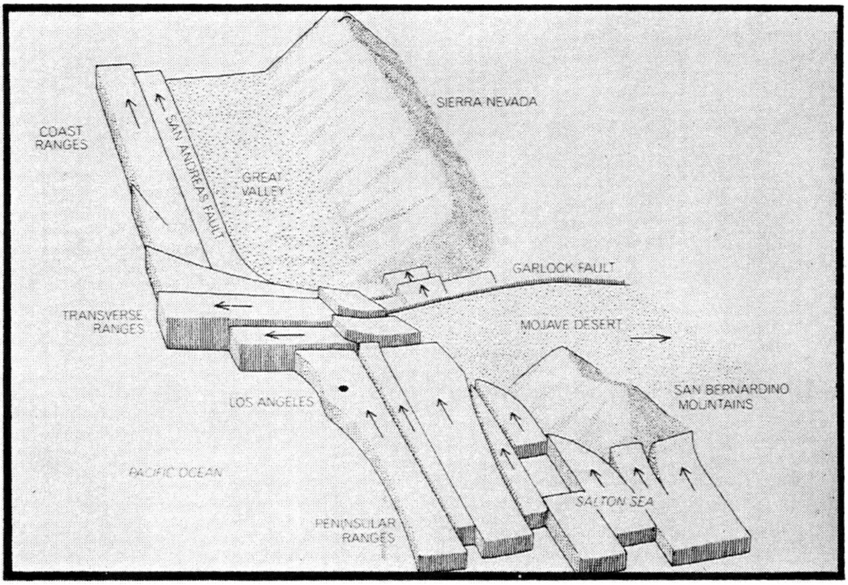

FIGURE 1-6 The San Andreas fault system.

In California prior to 1906, there was a rash of magnitude 6 and 6.5 earthquakes. Charles Richter, in his book on elementary seismology, states that there were approximately one dozen earthquakes of magnitude 6 to 6.5 that occurred in the 70 years before the 1906 earthquake. After the 1906 earthquake, there was a quiescent period where the rate of earthquake activity was not as high. This indicates that a build-up of stresses increased the tension within the crust of the earth, culminating in the 1906 earthquake, which was followed by a relaxation of tensions afterward.

Many seismologists and geologists feel that we are now entering a period of increased stresses and that we are once again going through a cycle of high earthquake activity. Whether that cycle will culminate in a magnitude 8 earthquake in 50 years, 100 years, or 200 years is a research question, but the concept of the cycle is fairly well accepted.

The following section focuses on the Loma Prieta earthquake and uses that event to illustrate some of the points about the effects of earthquakes.

The Loma Prieta earthquake occurred along the San Andreas fault system in the Santa Cruz Mountains (Figure 1-7). Generally, the earthquake was caused by the west side of the San Andreas fault moving upward about

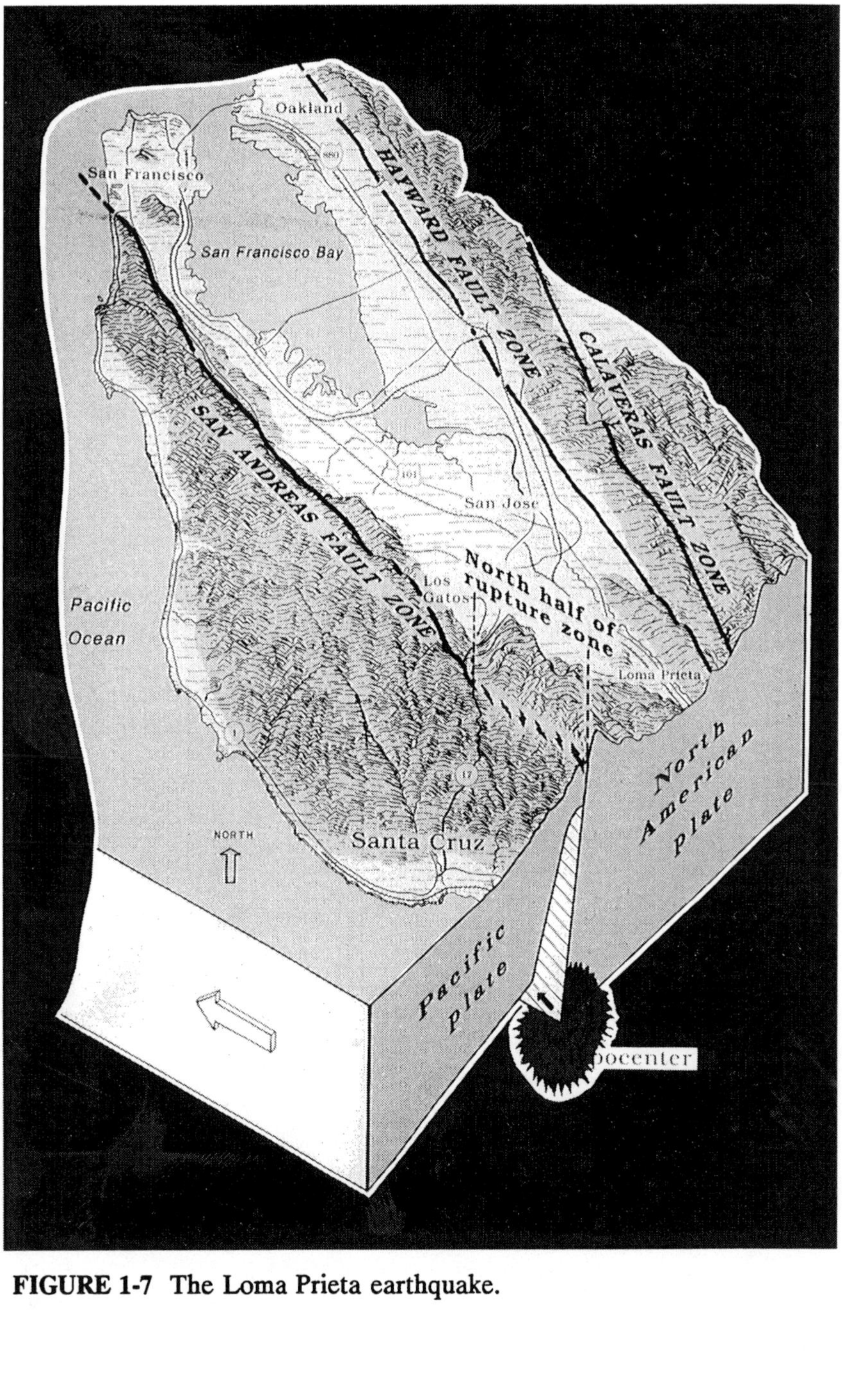

FIGURE 1-7 The Loma Prieta earthquake.

4 or 5 ft and shifting to the northwest about 4 or 5 ft. There was a sudden shift of the crust of the earth, up and to the northwest relative to the eastern side of the fault. This happened virtually instantaneously.

The effects of the earthquake were felt widely. Many of the effects occurred in the San Francisco Bay area, over 50 mi from the epicenter of the earthquake. This was not a big earthquake, and it was not located in the Bay Area. These are very important considerations because, had it been bigger and had it been in the Bay Area, the effects would have been much greater. Other effects were located in the Salinas River valley. Santa Cruz was hard hit. However, a lot of the damage was in the San Francisco area, particularly the Marina District and, of course, in Oakland, where the freeway collapsed.

Figure 1-7 shows a cross section of the San Andreas fault zone from the side, and from the surface to a depth of about 20 km (about 12 mi). The main shock occurred about 10 mi deep, and the aftershocks of the earthquake filled in a seismic gap along the San Andreas fault. The Loma Prieta earthquake, in effect, filled in a patch of the fault that had been stuck; that is, that had not moved in a long time.

Figure 1-8 illustrates cross sections of seismic activity along the San Andreas fault. These cross sections run from near San Francisco to central California. The top portion of the figure shows earthquake activity that had occurred in the several years before the Loma Prieta earthquake. In general, the fault is well covered with earthquake activity except for a gap where there have not been any earthquakes. Such a gap is, in effect, a stuck area on the San Andreas fault. The activity from the Loma Prieta earthquake which filled in the gap. Thus, we have developed a detailed understanding of the mechanics of how the San Andreas fault works, and we can identify other places along the fault that are stuck. This is one of the ways in which long-term earthquake prediction on the San Andreas fault can be carried out.

Figure 1-9 shows that the Loma Prieta earthquake occurred at a depth of about 16 km, or 10 mi. The fault break extended from that depth to near the surface, close to the San Andreas fault.

One of the important things about this earthquake, particularly with regard to economic consequences and the probabilities of earthquakes, is that this section of the San Andreas fault had been identified in a 1988 USGS report as the most likely place in northern California along the San Andreas fault for earthquake activity. Figure 1-10 was taken from that report. In northern California, there is an area in the southern Santa Cruz Mountains that approximately coincides with the section of the fault that subsequently broke. Therefore, this is encouraging to geologists and seismologists working on this problem, because it indicates that those places that are most vulnerable can begin to be identified. As a basis for allocating scarce resources in dealing with the earthquake threat, one can identify the most vulnerable places and focus efforts there. In southern California, there also are areas of great concern.

Following are a few of the effects of the earthquake. The Bay Bridge collapse (Figure 1-11), resulted from a flexing or a stretching of the bridge

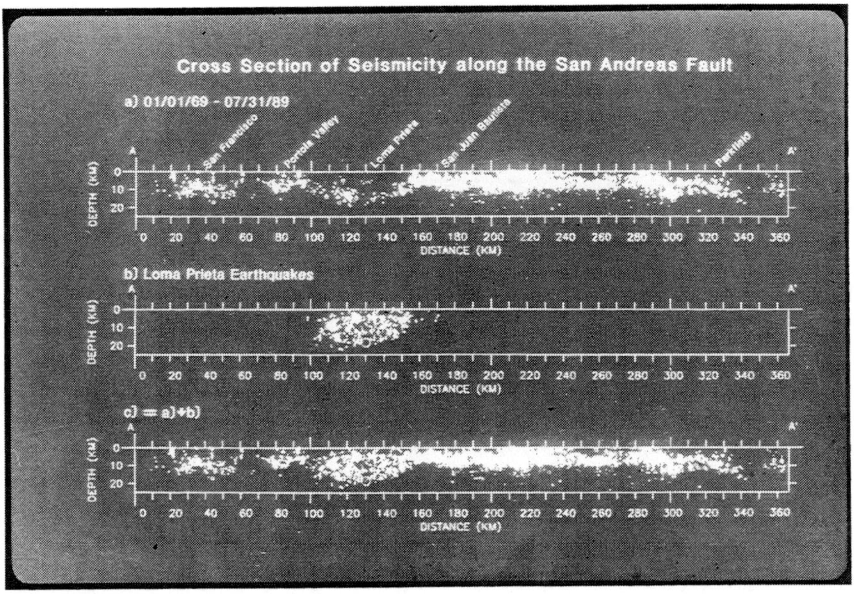

FIGURE 1-8 Cross section of seismicity along the San Andreas fault.

caused by the seismic waves. A very common problem in the Marina District was the failure of the ground and intense shaking. Numerous multistory buildings either collapsed or tilted quite a bit (Figure 1-12). Many of the problems were caused by open bays in the bottom of the buildings that are used as parking garages. There are not very many shear walls in the lower part of these structures, which allows them to shift.

In the Marina District in San Francisco, a phenomenon occurred that is very important. It is called liquefaction. Liquefaction occurs when sand is shaken and, in effect, turns into fluid that can then squirt out onto the surface of the ground, as in a volcano. Figure 1-13 shows evidence of liquefaction in the Marina. This is probably one of the phenomena that caused the intense shaking and ground failure in the Marina District. It is also important in other parts of the country. In the New Madrid area, in 1811 and 1812, thousands of square miles of land were covered with sand from intense liquefaction. We can anticipate that in the Mississippi Valley area, the next time there is a magnitude 8 earthquake, there will be a lot of water standing on the surface of the ground and a lot of sand deposited all over the place.

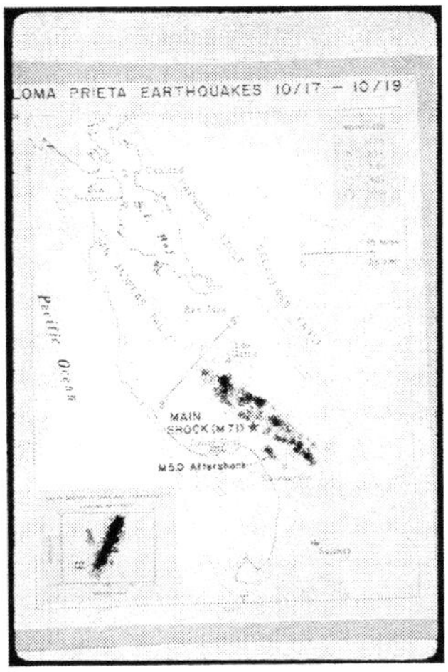

FIGURE 1-9 Loma Prieta earthquakes (October 17-19, 1989).

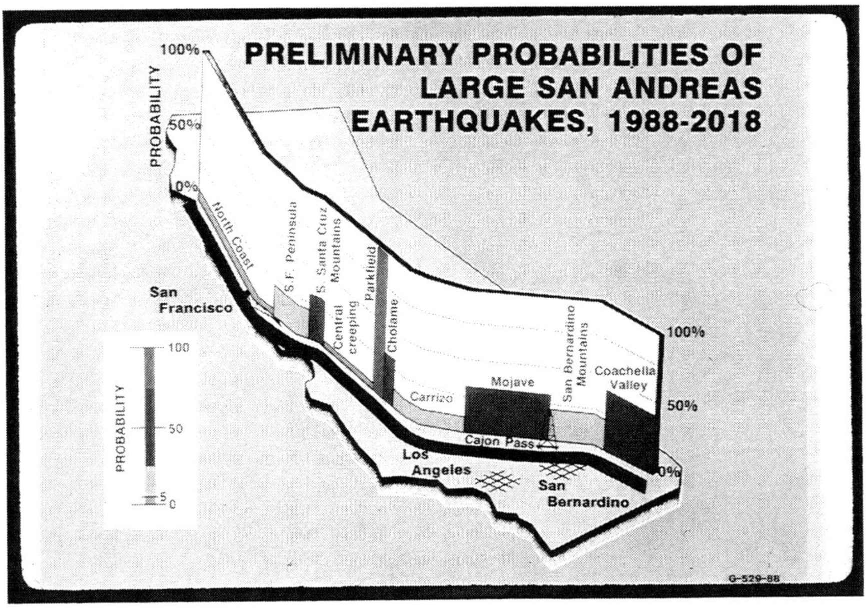

FIGURE 1-10 Preliminary probabilities of large San Andreas earthquakes (1988-2018).

FIGURE 1-11 Collapsed section of the Bay Bridge.

FIGURE 1-12 Multistory building in the Marina District.

FIGURE 1-13 Liquefaction in the Marina District.

Liquefaction and shaking played a role in the collapse of the bridge in the Salinas River valley (Figure 1-14). The pilings of this bridge had some problems at the top and also some movement at the base. Figure 1-15 shows the second deck and the bottom deck of the multistory Nimitz Freeway and illustrates the collapse of the pilings that held up the second deck.

In the Santa Cruz Mountains area, landslides and slumping were very common. Chasms formed in the ground (Figure 1-16), and in some places shaking was intense enough or ground failure occurred to cause the collapse of structures (Figure 1-17). Generally, wooden-frame houses do fairly well in the event of earthquakes. Usually unreinforced masonry structures have the most difficulties.

The areas of damage in the Loma Prieta earthquake in the San Francisco Bay area were largely predicted. Figure 1-18 illustrates the areas where maximum earthquake intensity was predicted. As shown, the Marina District, where a lot of the Loma Prieta damage occurred, was identified as an area of predicted violent shaking. This map replicates the effects that were seen in the 1906 San Francisco earthquake. It is quite possible for engineers and geologists to identify where ground shaking will be most intense from earthquakes, which is generally on unconsolidated sand or mud. Bedrock is a better place to be in the event of an earthquake. The importance of Figure 1-18 is that it shows that it is possible to map an area based on engineering and geologic studies in order to identify the most vulnerable areas.

FIGURE 1-14 Collapsed bridge in Salinas River valley.

One area of concern in California is the Hayward fault in the Bay Area. The USGS recently reanalyzed the threat of earthquakes along that segment and have raised the probabilities of earthquakes along the Hayward fault. A lot of critical structures are located along the Hayward fault. Schools are nearby, and lifelines cross this area; therefore, it will require attention in the next several years.

Focusing on the earthquake history in the rest of the country, Figure 1-19 shows earthquakes of intensity 7 or greater that are known in historical time. They are fairly broadly distributed; however, there are concentrations. One of the biggest concentrations is in the central Mississippi Valley area, in the area of the New Madrid seismic zone, which has been--over the last several hundred years and even recently--the most active seismic area in the eastern United States. There has also been a lot of activity in the St. Lawrence River valley area.

As recently as the last 15 years, there was no geologic explanation for the cause of the New Madrid earthquakes. But now, thanks largely to the research that has been carried out under the National Earthquake Hazard-

FIGURE 1-15 Nimitz Freeway, showing collapsed pilings.

FIGURE 1-16 Chasms in the Santa Cruz Mountains.

FIGURE 1-17 Collapsed house at Boulder Creek.

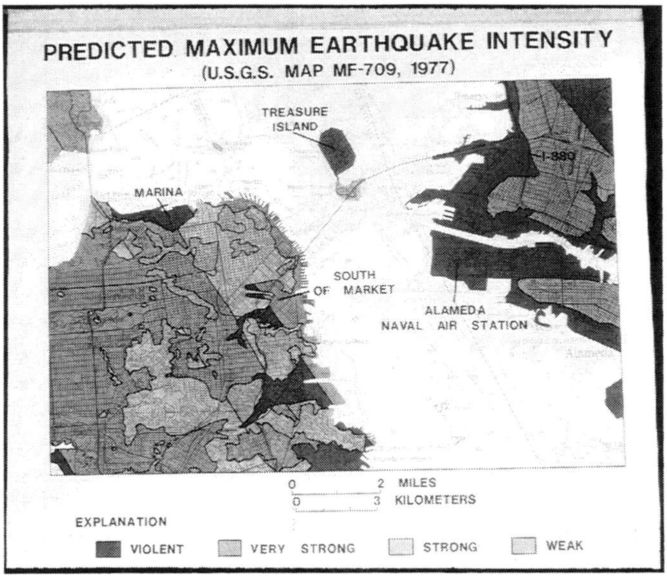

FIGURE 1-18 San Francisco Bay area: predicted maximum earthquake intensity.

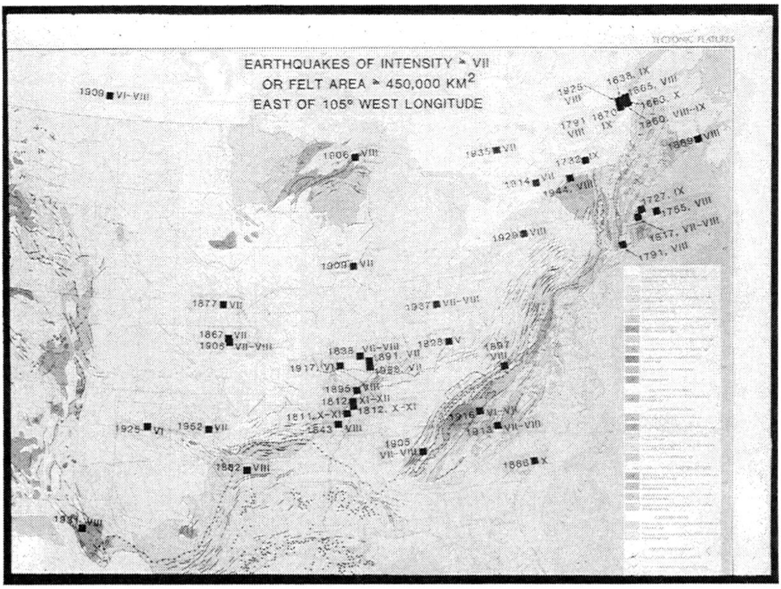

FIGURE 1-19 Earthquakes of intensity ≥ VII or felt area ≥ 450,000 km^2.

Reduction Program, we have a fairly good understanding of the cause of these earthquakes.

As stated at the beginning of this chapter, earthquakes in the East affect a far greater area than earthquakes of corresponding magnitude in the West. Figure 1-20 shows the areas of intensities 6 and 7 for the New Madrid earthquake of 1811, the Charleston earthquake of 1886, the San Francisco earthquake of 1906, and the San Fernando earthquake of 1971. The 1906 earthquake had a magnitude of about 8.25, and this particular New Madrid earthquake also had a magnitude of near 8, indicating that about the same amount of energy was released. But it is clear how much greater an area was affected by the eastern earthquake. The reason for this is the crust of the earth in the East is older and colder and less fractured, or at least the fractures are healed somewhat relative to those in the West. The West is young and active and dynamic--referring to geology, not people--so seismic waves are more attenuated in the West compared with the East. If you ring the crust in the East, the waves propagate much more efficiently and to a greater distance than in the West. The same is true of the Charleston earthquake, which was on the order of magnitude 7, and the San Fernando earthquake, which was more like 6.5; but the Charleston earthquake affected a much larger area. This means that although earthquakes occur less often in the East, they affect a greater area when they do occur.

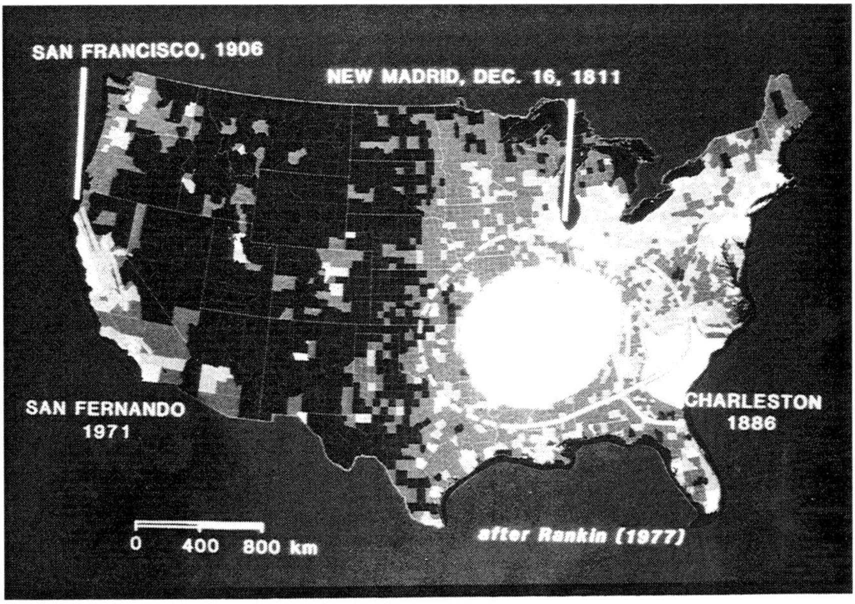

FIGURE 1-20 Felt areas of some large U.S. earthquakes.

Figure 1-21 shows the terrain (elevations) in the eastern United States. There is a depression in the crust of the earth that extends into southern Illinois and Indiana, which is anomalous in the East.

Figure 1-22 illustrates the attraction of gravity in the East. There is a high gravitational attraction relative to elsewhere, extending into the central Mississippi area, which also indicates a somewhat anomalous crust of the earth in the Mississippi Valley area. The earthquake activity in the New Madrid area occurs along some fairly narrow boundaries running through northeastern Arkansas, into the Boot Heel of Missouri.

The bands of seismic activity shown in Figure 1-23 define faults that are associated with specific geologic features. In addition to the earthquakes that occurred in 1811 and 1812 in this area, there also have been other earthquakes since then. In 1843 in northeastern Arkansas, there was a magnitude 6.5 earthquake; and in 1895, there was a magnitude 6.8 earthquake (very close in magnitude to the Loma Prieta earthquake) in the area of Charleston, Missouri. Since 1895 there has not been a 6 or greater magnitude earthquake in this particular area. There has been a long period of time without significant earthquake activity in the New Madrid zone, but it is still active. It just has not "popped off" lately.

Figure 1-24 is a map of the magnetic field of the New Madrid region. A band of subdued magnetic field runs along this trend. The earthquake

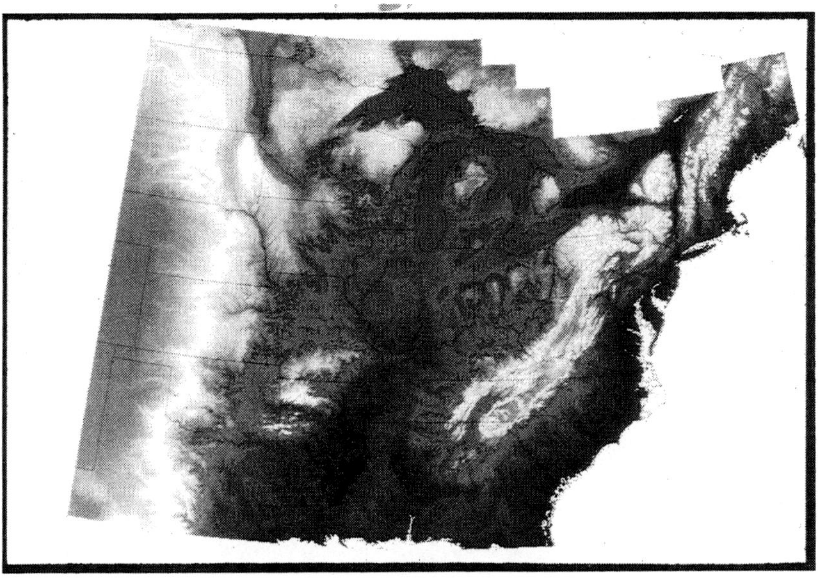

FIGURE 1-21 Terrain map of the eastern United States.

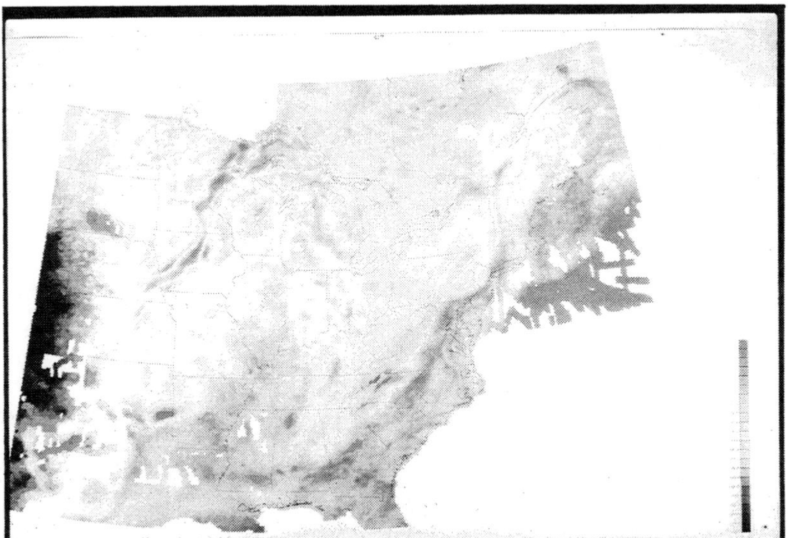

FIGURE 1-22 Gravity map of the eastern United States.

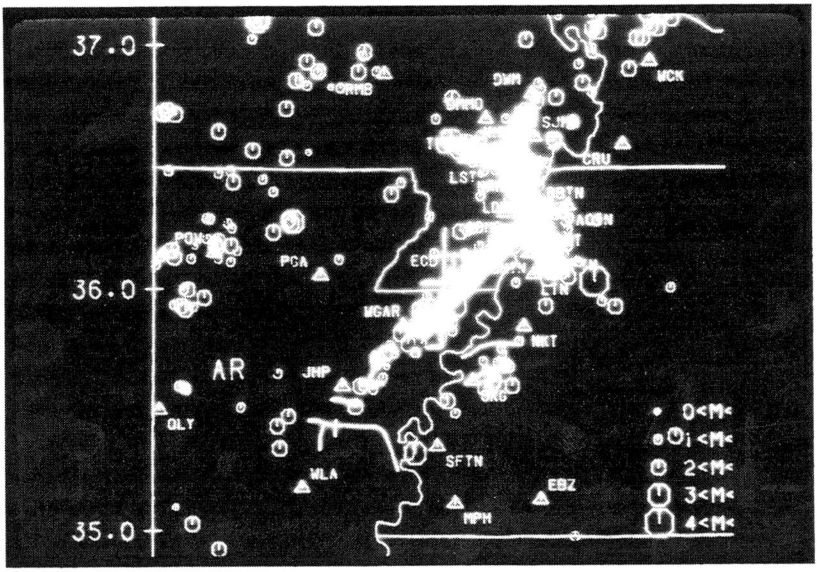

FIGURE 1-23 Seismic activity in the New Madrid region.

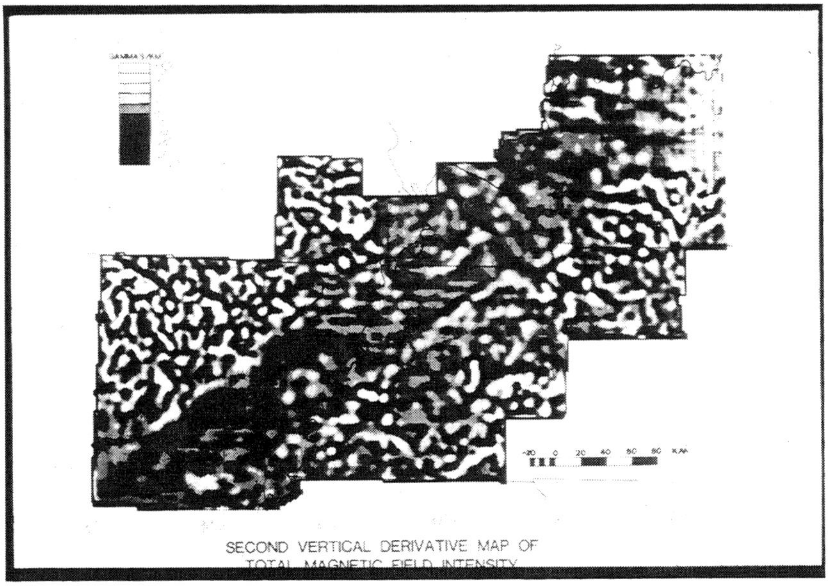

FIGURE 1-24 Magnetic field in the New Madrid region.

activity runs right down the middle of the band, then crosses it and runs along its northwestern boundary. This band of activity is caused by a down-drop of about a mile of the crust of the earth. It is much like a keystone in an arch: If you stretch the arch a little bit, the keystone drops down. In this case, the keystone dropped about 500 million years ago when the crust of the earth in this area was stretched. Since then, the stresses that have been caused by the flexing of the North American plate have repeatedly jostled the keystone and created a new geologic structure, but it is basically the reactivation of the keystone that is causing earthquakes now.

In 1886, a magnitude approximately 7 earthquake occurred in the Charleston, South Carolina, area and caused significant damage. It also caused liquefaction, land deformation--including a deformation that bent railroad tracks--and fairly extensive sand-blow activity. These blows are very useful, because they show evidence of prehistoric earthquake activity. These are very common in the New Madrid area, and more of them have recently been found in the Charleston area. Also some were found recently in the Wabash Valley area between Indiana and Illinois, which indicates earthquake activity in that area.

The earthquakes along the eastern margin of the United States are very poorly understood, but they probably have to do with the formation of the Atlantic Ocean. There is a line of earthquake activity that runs right down the middle of the Atlantic Ocean. In the past, the continents were joined together. Because of hot material coming up from the inside of the earth, the continents drifted apart to form the Atlantic Ocean. The scars that were left when those continents drifted apart probably established the fault lines that are being reactivated now to cause the earthquakes along the Atlantic margin.

When it comes to assessing the economic impact of earthquakes and designing various measures for earthquake mitigation, it is very important to know about earthquake probabilities. These probabilities can be estimated from geologic and seismologic evidence. This evidence has been referred to at various locations in this chapter. In California, that type of data is in fairly good abundance. Estimates have been made of the likelihood of earthquakes along various parts of the San Andreas fault; these estimates have been published. These probabilities are of the nature of a 30 percent chance of a certain-sized earthquake occurring in the next 30 years.

Studies on the economic impact of earthquakes in the eastern United States and the midcontinent, are much further behind. There have been estimates based on fairly fragmentary data. There are estimates for the New Madrid area, giving numbers like a magnitude 8 earthquake every 1,000 years or so. But these estimates are very unreliable at the present time. Also, there must be some geologic phenomena there that can be found to improve those estimates.

Overall, over the next 10 to 20 years, there will be new data that will greatly improve our estimates of earthquake likelihood. These data can be used, together with engineering and other types of efforts, to greatly reduce the vulnerability of earthquake damage and the threat to the country from earthquakes.

GENERAL DISCUSSION OF CHAPTER 1

QUESTION: When the New Madrid zone is projected to the northeast, there is a basis for projecting it all the way to the Buffalo area. What is the likelihood of earthquakes further north?

DR. HAMILTON: There is a hypothesis that the earthquake activity in the St. Lawrence Valley area in the northeast could be projected all the way down to the New Madrid area. There is a rough alignment of epicenters along that trend. So when a seismologist looks at the dots on an epicenter map, it is pretty easy to draw a line through them. Geologists, however, get very agitated when that is done because they look at the geologic structures and say the continuity in the structures is just not there. Therefore, it has to be regarded as an unproven hypothesis.

Even looking in the New Madrid area, geologically there is no continuity in the faults between the Wabash Valley area and the Mississippi Embayment area. All of the faults that have been mapped terminate. There is a discontinuity along another fault system that runs east-west, so it is even conjectural whether the structures in northeastern Arkansas can be connected with those between Indiana and Illinois. Lacking a convincing case that these structures can be connected, a reliable projection of what significance New Madrid activity might have further to the northeast cannot be made.

QUESTION: What is the story behind the predictions of the Loma Prieta earthquake?

DR. HAMILTON: In the sense that the term "prediction," is normally considered, that earthquake was not predicted. There was a report published in 1988, the year before the earthquake, that identified that section of the fault as having a high probability for future earthquake activity. A probability of 30 percent was given for an earthquake in the next 20 years, but in no sense was that section identified as having a high likelihood of an earthquake in the near term. Therefore, it would be what one might call a long-term earthquake prediction.

The evidence for that prediction was largely geologic and seismologic in that it was a place where there was a gap in the seismicity. This was a case where the stresses were building up and were not being released. Also, based on geologic and surveying data, scientists knew that there was a slip deficit that sooner or later had to go. It was a case where the area was identified as a likely place for an earthquake, but there was no near-term prediction.

After the earthquake, there was some evidence that suggested that maybe there were some phenomena beforehand. An instrument run by a professor at Stanford, which monitored electromagnetic phenomena, had recorded anomalous activity a few hours and days before the earthquake. But, there is no reliable basis for a short-term earthquake prediction. However, there is a good basis for identifying potentially dangerous areas based on longer-term studies.

QUESTION: Has there been a probability associated with a New Madrid earthquake?

DR. HAMILTON: There have been some recurrence estimates given, and there have been some probabilities published, but they are not regarded as being very reliable. Generally, a magnitude 6 earthquake in that part of the country should be expected every several tens of years or so. A magnitude 8 earthquake might occur every 500 to 1,000 years.

QUESTION: Figure 1-18 showed the high probability of earthquake likelihood on the east side of the San Francisco Bay. Do those areas also indicate something about damage?

DR. HAMILTON: Yes. Figure 1-18 was a map of predicted violent ground shaking that the USGS published in 1977 and illustrated that there was a lot of damage in the Marina area. Also, the area where the Nimitz Freeway collapsed had been identified as an area of poorly consolidated material. Of course, there was not a lot of damage in the East Bay, but this earthquake was a long way from the East Bay. If an earthquake occurred in the Bay Area, particularly along the Hayward fault, a lot of damage in areas of poorly consolidated material could be expected.

2

What are Likely Categories of Loss and Damage?

Chapter 2 provides a basis for understanding how loss estimates are generated for different categories of losses and damages. Specific issues that will be addressed include: What loss-estimation methodologies currently exist? How satisfactory are inventories for these approaches? In relation to actual losses, how good are these estimates? Are there methodologies for estimating nonstructural losses? What data bases are available from which projections of nonstructural losses can be made?

Mr. Christopher Arnold is an architect and the president of Building Systems Development in California. He is a Fellow of the American Institute of Architects, elected for his contributions to research in the architectural aspects of seismic design. Mr. Arnold is also a member of the NAS/NRC Committee on Earthquake Engineering and has served on that committee's panel on loss-estimation methodologies. Mr. Arnold will present an overview of loss-estimation approaches.

Dr. Don G. Friedman is the director of the Natural Hazards Research Program at the Travelers Insurance Company. Dr. Friedman has 35 years of experience in the assessment of casualty and damage potentials of natural disasters for the insurance industry, federal agencies, and most recently for the All-Industry Research Advisory Council. His presentation focuses on risk-assessment models and the types of data necessary to estimate casualty and property loss potentials from a catastrophic earthquake.

Professor Kathleen Tierney is an associate professor of sociology and the research director of the Disaster Research Center at the University of Delaware. She is the author of a number of monographs, articles, and book chapters focusing on various hazard- and disaster-related topics, including socioeconomic consequences of earthquakes. She is a member of the Advisory Committee for the National Earthquake Hazard-Reduction Program. The topic of Professor Tierney's presentation is on loss estimation and public policy from a social science perspective.

Professor Robert W. Kling is from Colorado State University. Dr. Kling has a doctoral degree in economics from the University of Kansas and has recently been involved in three National Science Foundation (NSF) projects that have addressed different aspects of social and economic effects of different types of natural hazards. The result of one of these projects is a work entitled *Natural Hazards Damage Handbook: A Guide to the Uniform Definition, Identification, and Measurement of Damages from Natural Hazard Events*. In his presentation, Dr. Kling focuses on the loss of the cultural environment from a catastrophic earthquake.

PRESENTATION OF CHRISTOPHER ARNOLD

This section provides an outline of the methods that are currently used for developing estimates of loss, and an outline of the nature of loss focusing on one particular building in the Loma Prieta earthquake. The information on loss-estimation methods is based on the Academy panel study on estimating losses from future earthquakes conducted under the chairmanship of Bob Whitman a few years ago. But the following information does not necessarily represent the views of the panel or the Academy.

The typical parameters of a loss-estimation study shows that the following things have to be determined before developing a specific study:

- the type of loss (e.g., casualties, the number of homeless, the functionality of essential facilities, and economic impact);
- the kinds of facilities (i.e., all buildings or structures and essential facilities like hospitals and lifelines);
- the degree of certainty, and some feel for the degree of detail which is necessary--these obviously have very large cost and time implications; and
- the time span and the kind of seismic risk.

Perhaps the number of earthquakes which might occur in a time span is of particular interest. Either a predicted earthquake or an actual historic earthquake may be used. Studies have been done, for instance, that show the impact of the 1923 Kwanto earthquake on today's Tokyo. Also, before developing a loss-estimation study, questions of geographic scale--local, regional, state, or national--must be addressed.

Typically, the kind of loss-estimation studies that have been done, other than those which may have been done for very specific purposes such as the inventory of a large corporation's set of buildings, are studies such as the NOAA studies of the San Francisco Bay area in the 1970s; the FEMA studies in selected cities, such as St. Louis and Boston; or the midwestern six-city study. These have provided damage estimates that are expressed in dollar terms and estimates of casualties. They tended to deal with all buildings, although some of them have focused somewhat on essential facilities such as hospitals. The degree of certainty is probably low. The degree of detail, because of the cost of producing the study, is also low. The estimates of seismic risk typically use the modified Mercalli scale, for better or for worse. It is at least generally understood and accepted, and has been commonly used as a way of defining the seismic risk. These studies typically have been at a regional or large-city level.

The use of these studies has primarily been political. They have been used to assist the politics of earthquake-hazard mitigation and the process of consciousness raising, and they have also been used to support some of the commercial aspects of the "earthquake industry," under the new circumstances by which the earthquake problem is starting to become a recognized industry.

Recognizing that these studies generally have a large amount of scientific qualification on the way in which they have been done and the authority of the statements, the political use of these studies has been to look for the "large numbers" and to use those to attempt to increase earthquake awareness and hazard-mitigation activities.

The methodology of loss estimation is extremely simple and obvious. Some time ago Mr. Arnold wrote a paper on this, which stated that methodologies are cheap, but data is very, very expensive. All studies follow the same basic methodology:

- develop an inventory of building stock;
- define the seismic risk that applies to those buildings; and
- utilize a mechanism for relating motion, damage, and loss so that damage to the inventory can be estimated.

This is then typically converted into the dollar loss that is received from exposure to a given seismic risk, as applied to that inventory, and out of that comes a loss estimate definition. That is the general methodology. To focus on one particular aspect of it, the inventory and the motion-damage-loss mechanism must use the same building classification, therefore a classification for defining the inventory of buildings must be developed that has to be the same classification used to apply the motion-damage-loss mechanism to the inventory. That sounds obvious and simple but, in fact, it has proven to be rather difficult.

Table 2-1 shows a classification system for buildings which is very widely used; it was developed originally by Karl Steinbrugge for the Insurance Service Offices in the Bay Area and has been slightly modified since. It is a simple and broad classification, with 21 different building types. One who is unfamiliar with buildings may think that is a lot. In fact, it is a very small classification because when a building inventory is inspected, every building must be assigned to one of these 21 types. Within each type there will be a great deal of variation; that presents difficulties, but nevertheless this is a very widely used system.

Another system was developed somewhat later under a program called ATC-13 (Applied Technology Council Study Number 13). This was a FEMA-sponsored study which was intended to bring the loss-estimation technique to a more advanced level. It was also intended originally to go right through to loss-estimation studies which would be used to estimate economic losses, industrial capacity losses, and so on. That was never quite achieved, but a lot of the study intent was accomplished. The study uses a rather more complete classification system than the Insurance Services method. The classification system has 40 building types instead of 21, so that it is a slightly finer-grained classification system.

For any study, the inventory is critical, because this is the whole basis upon which you are going to assign your loss estimation. Unlike other aspects of loss estimation, an actual inventory exists. In other words, there is a finite

TABLE 2-1 Construction Classes Used in the ISO and NOAA/USGS Methods

Building Class	Brief Description of Building Subclasses
1A-1	Wood and stuccoed frame dwellings regardless of area and height
1A-2	Wood and stuccoed frame buildings, other than dwellings not exceeding three stories in height or 3,000 square feet in ground floor area
1A-3	Wood and stuccoed frame structures not exceeding three stories in height regardless of area
1B	Wood frame and stuccoed frame buildings not qualifying under class 1A
2A	One-story, all metal; floor area less than 20,000 square feet
2B	All metal buildings not under 2A
3A	Steel frame, superior damage control features
3B	Steel frame, ordinary damage control features
3C	Steel frame, intermediate damage control features (between 3A and 3B)
3D	Steel frame, floors and roofs not concrete
4A	Reinforced concrete, superior damage control features
4B	Reinforced concrete, ordinary damage control features
4C	Reinforced concrete, intermediate damage control features (between 4A and 4B)
4D	Reinforced concrete, precast reinforced concrete, lift slab
4E	Reinforced concrete, floors and roofs not concrete
5A	Mixed construction, small buildings and dwellings
5B	Mixed construction, superior damage control features
5C	Mixed construction, ordinary damage control features
5D	Mixed construction, intermediate damage control features
5E	Mixed construction, unreinforced masonry
6	Buildings specifically designed to be earthquake resistant

number of buildings of certain kinds. The problem is that it almost never exists in any published form, and the costs of achieving that are astronomical; when one talks about defining an inventory, he is really defining some kind of simulation or subterfuge for the actual inventory that exists.

The Academy panel spent 2 days discussing this inventory question: how is it determined? and how is an inventory defined? For instance, the census does not reveal the things about buildings needed to determine loss estimation. Assessor's records do not indicate the things needed to know about loss estimation. The top half of Figure 2-1 shows an idealized version of achieving

FIGURE 2-1 Earthquake-damage-loss estimation.

an inventory, in which the buildings would have some sort of computerized system that would send out their accurate vital statistics. The lower picture in Figure 2-1 shows the way inventories are actually done: A small number of people gather together in a room and assign buildings to their correct classification. You will find that the smoke-filled room reoccurs rather often in the loss-estimation methodology.

The loss and damage mechanism, developed primarily by Karl Steinbrugge, is critical and is shown in Figure 2-2. It is simple and relatively easy to apply. The designations of the building classification system are shown as the curves: the modified Mercalli intensity is on one axis and the per-

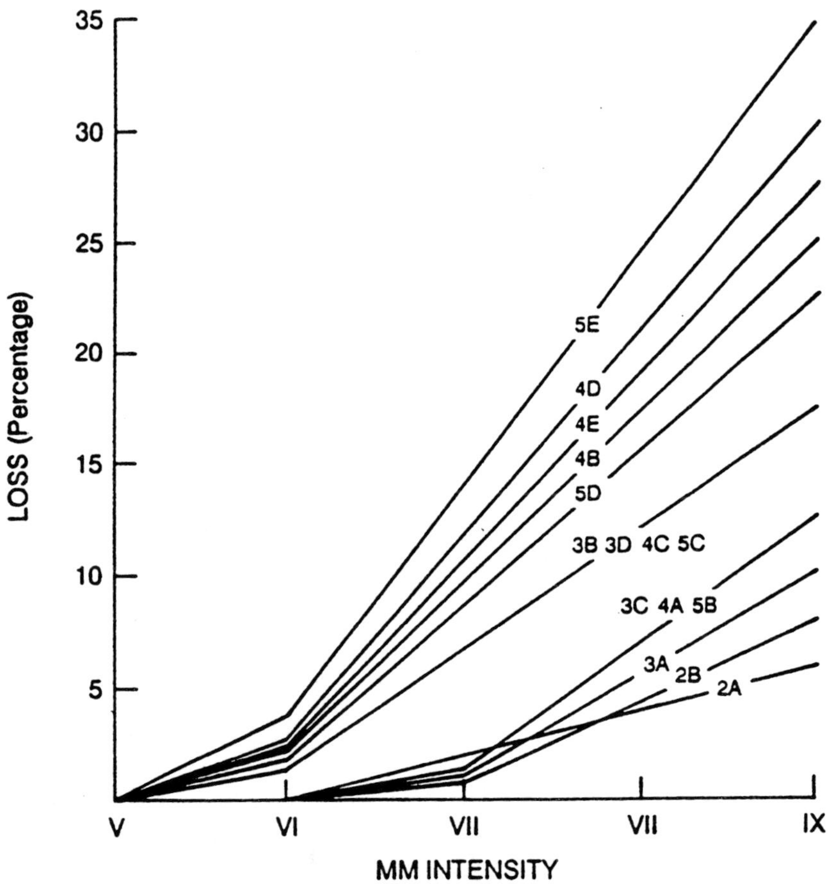

FIGURE 2-2 Loss ratio versus modified Mercalli intensity (mean damage ratio curves). FROM: Estimating Losses from Future Earthquakes, p. 29, National Research Council, 1989; SOURCE: Algermissen and Steinbrugge, 1984.

centage loss is on the other axis. Select a given type of building, such as unreinforced masonry, and find the curve for it, then find the modified Mercalli intensity, and establish a percentage loss. These are obviously very gross figures; as they are studied, all kinds of reasons why they may be inaccurate can be thought of. This, however, is currently the level at which these estimates are done.

The ATC-13 process developed a somewhat more elaborate methodology, and the team was also interested in making the methodology more transparent. Rather than the smoke-filled room, the idea was to have a methodology in which its method of development was clear and could continue as more information came in. The project developed a general matrix, called a *damage probability matrix*, which defines damage states in words. These are defined numerically and can in turn be developed into a dollar-loss estimate related to the modified Mercalli scale. The estimates, of course, vary according to the building type.

The actual estimates were developed by expert committees in a delphi process with a number of experts filling in forms and, in effect, voting on their estimates of the correct numbers. Figure 2-3 shows the scatter: for one particular class of building, the low estimate, the best estimate, and the high estimate. The symbols show the estimates of specific people. There were two rounds, with a fairly small number of people involved. The object of the two rounds was to try and reduce the aberrations in the estimating, although sometimes the aberrations may be correct and the "enforced" agreement may not be correct.

Nevertheless, the process arrives at a set of numbers that can be used in the actual matrices. Figure 2-4 shows the ATC-13 matrix for facility class 18, which is a low-rise, concrete, movement-resistant, frame-building type. It can be seen that the matrix shows fairly small amounts of damage at even the high modified Mercalli figures, and 100 percent damage would be expected in this class of building for any modified Mercalli figure. One can agree or disagree with that, but this procedure enables a dollar figure to be arrived at for a given building type related to a range of modified Mercalli figures.

There is another system which has been used. This is the "fragility curve," which was developed by a consultant and used for one of the FEMA studies (Figure 2-5). This is really a rearrangement of the same basic material, in which the peak ground acceleration or Mercalli intensity is used. The numbers 1, 2, 3, 4, and 5 represent different damage states. All these systems are pushing around the rather small amount of real data that there is about the effects of earthquakes on actual buildings.

The above is the essence of how loss estimating is done. The question then comes up as to how accurate is the information that is received from this sort of process. Figure 2-6 shows one of the studies that was done for the Academy report, in which the curves represent different estimates of damage to wood-frame buildings based on various research studies which people have done at different times. The black spots are actual recorded damage, so the range of variation between different consultants' estimates and how those

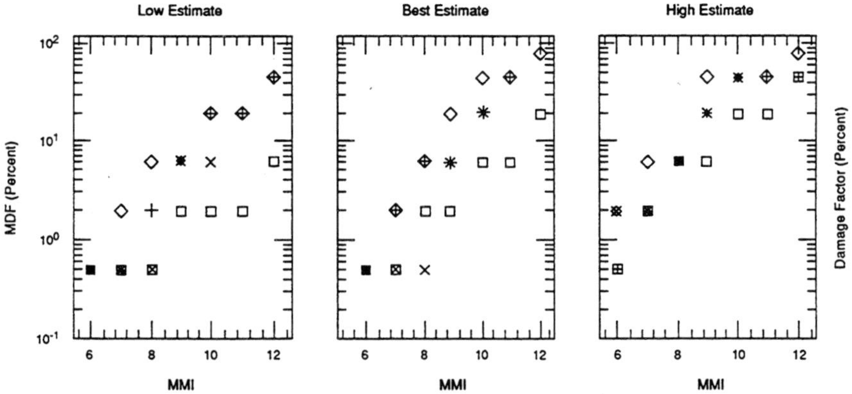

FIGURE 2-3 Expert responses to round one damage factor questionnaire for Facility Class 18--low-rise moment-resisting ductile concrete-frame buildings. NOTE: Each symbol represents the estimates of one specific person.

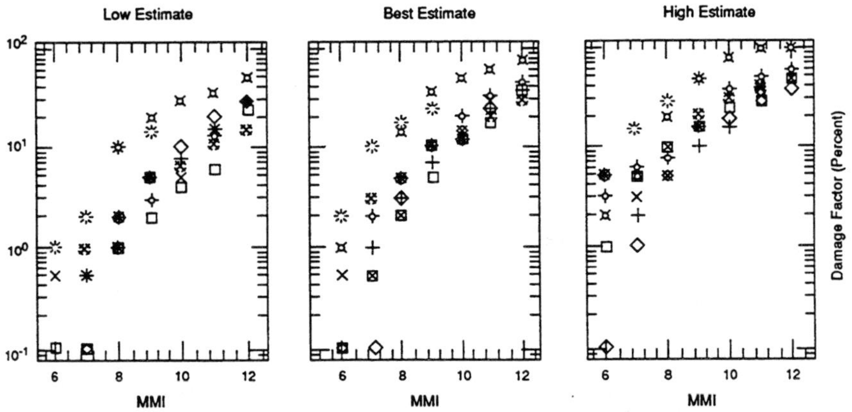

FIGURE 2-4 Expert responses to round two damage factor questionnaire for Facility Class 18--low-rise moment-resisting ductile concrete-frame buildings. NOTE: Each symbol represents the estimates of one specific person.

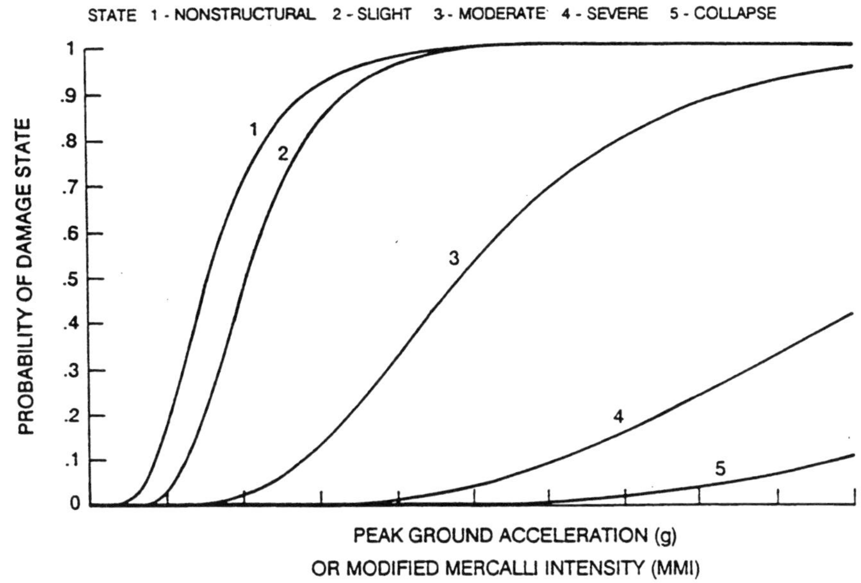

FIGURE 2-5 Fragility curves for wood-frame buildings. FROM: Estimating Losses from Future Earthquakes, p. 38, National Research Council, 1989; SOURCE: Kircher and McCann, 1983.

estimates relate to actual damage is clear. Not nearly enough of this sort of validation exercise is being done. So far, a number of loss estimates have been done, but very little in the way of validation. Validation is rather expensive and does not seem to have the sort of appeal to the research community that other aspects of the earthquake problem have, but it is very critical and very important. As earthquake occurrences, such as Whittier and Loma Prieta, continue more information is developed but methodologies are not being reviewed and validated like they should be.

A frequent subject of interest in loss estimation is deaths and injuries. This is perhaps even more vague than the dollar-loss aspect, but there is a table in the ATC-13 study which, again, was developed in a smoke-filled room by a small number of people (Table 2-2). By applying this table, depending on the damage state, some estimate of injuries and deaths can be calculated. This may not be very accurate, but it is certainly much more useful than just speculating about the number of injuries and deaths.

Some of the Academy committee members pushed to try and get some numerical estimate of accuracy. The engineers were rather reluctant to do this, but some numbers were published that are interesting. One was that estimates for single-family wood-frame houses, where there is a lot of experience, might perhaps be accurate to within a factor of about 1.5. For run-of-the-mill

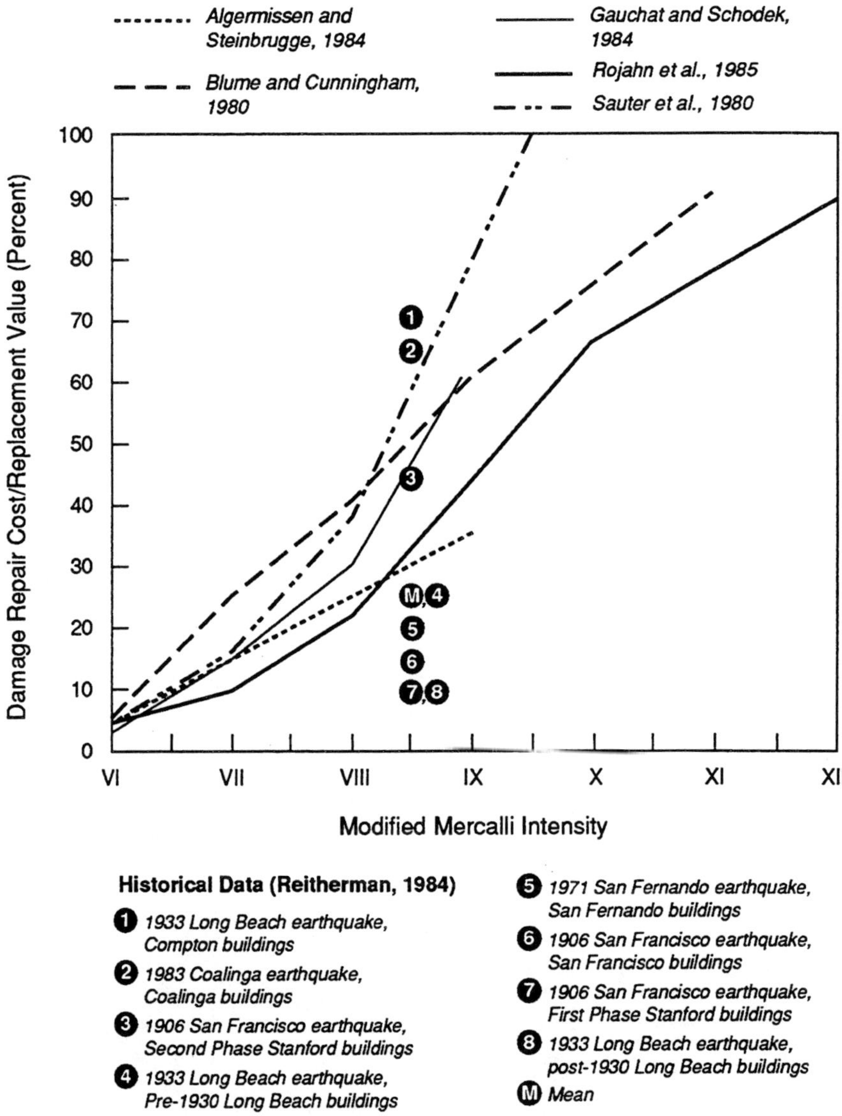

FIGURE 2-6 Intensity-damage relationships for unreinforced masonry buildings. SOURCE: R. Reitherman, 1988.

TABLE 2-2 Injury and Death Rates in Relation to Damage

Damage State	CDF(S) Percent	Fraction Injured Minor	Fraction Injured Serious	Fraction Dead
1	0.0	0	0	0
2	0.5	3/100,000	1/250,000	1/1,000,000
3	5.0	3/10,000	1/25,000	1/100,000
4	20.0	3/1,000	1/2,000	1/10,000
5	45.0	3/100	1/250	1/1,000
6	80.0	3/10	1/25	1/100
7	100.0	2/5	2/5	1/5

NOTE: Estimates are for all types of construction except light steel construction and wood-frame construction. For light steel construction and wood-frame construction, multiply all numerators by 0.1. SOURCE: Applied Technology Council, 1985

commercial-type buildings, the accuracy might be within a factor of 3, and for buildings in areas where there is not a lot of seismic activity, outside California, the accuracy was perhaps an order of magnitude--a factor of 10. In scientific terms this is very inaccurate. In other terms, however, it may be useful. It is useful to know whether there will be hundreds of houses down or thousands of houses down.

Table 2-3 shows the result of a study that was part of the Academy study, which compared the ATC-13 and the Steinbrugge figures. This shows that for wood-frame buildings, for instance, the ATC-13 study has a damage ratio of 8.8. The Steinbrugge study has a ratio of 8 or 12, depending on the kind of building, so that is fairly close. If one looks at tilt-up, ATC-13 shows a damage ratio of 16 percent; the Steinbrugge curve shows a damage ratio of 30. Again, depending on the viewpoint, this is a wild spread, or it is quite useful in terms of what it is used for.

A new development in loss estimation has been its entry into the commercial area. An example of a commercial project, to some extent sponsored by the insurance industry, and really directed at providing information of specific value to the insurance industry, is a project based on research done originally at Stanford University. The Insurance Investment Risk Assessment System (IRAS) Project is a computerized system in which for a given site or region damage to an individual building or an inventory of

TABLE 2-3 Comparison of Some Building Damage Ratios (D/R)

U.S. Geological Survey (USGS) and ATC-13 @ MM IX

ATC-13 Name	No.	D/R	USGS Code	D/R
wood frame	1	8.8	1A	12 old 8 new
light metal	2	5.6	2A small 2B large	6 8
VRM				
low	75	42.0	5E	35
medium	76	52.9		
braced steel frame				
medium	13	11.3	3A	10
high	14	14.0		
concrete DMRF				
low	18	8.7	4A	13
medium	19	10.3		
high	20	12.5		
Tilt-up	21	15.8	4D	30

SOURCE: H. Degenkolb

buildings can be estimated and the cost/benefit of retrofitting can be provided. Losses can be related to any valid earthquake expressed as a maximum risk or averaged to determine the probable expected losses in a given time frame.

This type of system is a new development which represents a new evolution. Such systems still essentially follow the same methodological basis and use the same information, but output is provided in a much more usable form. At the moment, this particular system applies only to California, in which the hazard is fairly well defined.

What, in fact, is a loss? These studies end with a dollar figure which represents the damage to the building, converted into a dollar figure for replacement. But, there is much more to it than that. The loss does not stop at the damage.

The Hyatt Regency Hotel in Burlingame, near San Francisco Airport, is a new, 400-bed hotel that was completed about a year before the Loma Prieta earthquake, and is a 400-bed hotel. It is a rather nice building architecturally; like most of the Hyatt Regency Hotels, it is built around an atrium, which is a 10-story lobby. It has an interesting translucent fabric roof, which is supported on light steel frames. The restaurants and bars in the center of the atrium, underneath which are the main meeting areas within the hotel, can be seen looking down from the galleries. It is an interesting architectural plan.

The hotel suffered some damage in the Loma Prieta earthquake. It suffered some nonstructural damage, but this was symptomatic of some structural damage within the building. This building may be subject to litigation, so little information is available. The following summary is based on published information, which may or may not be correct. The structural damage consisted of some severe but repairable damage to concrete beams and shear walls in the lower portions of the building.

The hotel was fully occupied, and nobody was killed or injured, but it was decided that the damage was severe enough to justify closing the building, so the building was closed, repaired, and reopened on July 20, 1989, nine months after the earthquake.

The loss has been quoted as $12 million in damage repair costs and $12 million in lost revenues between October and July, while the hotel was closed. This might be described as a $12 million direct loss and a $12 million indirect loss. The cost of construction was about $54 million, so perhaps the direct loss was on the order of 25 percent (which probably would not have shown up using typical loss-estimation methods).

Because the loss to the Hyatt Company was insured, the loss was to the insurance company. The building was rebuilt, so this represented a gain to the architects and engineers; the figure of $5 million in fees has been quoted, though this may not be true. Two separate engineering firms have been involved in the hotel's rehabilitation work. There have been arguments and controversy, so perhaps $5 million may be correct. The building contractors have gained money, and the local construction labor force in Burlingame has gained money. Attorneys have gained, to date, something on the order of a few hundred thousand dollars. In fact, this situation represents a new redevelopment project, an unexpected redevelopment project for the contracting industry in that area.

Some people lose, some people gain. It could be argued that the insurance companies have not lost. The insurance companies are simply providing their services. When an architect designs a building for someone, the work should not be regarded as a loss; it is a service that is the architect's job to provide. It is only a loss if an error is made in calculating what it will cost to provide the plans. Thus, if the insurance companies make up Hyatt's loss, they are simply providing a similar service. The Hyatt company obviously had not paid an equivalent sum in premiums in the year in which the hotel existed, but the insurance industry is not based on that sort of arithmetic.

Hyatt's loss of revenue was balanced out by other airport hotels, which gained revenue. It has been quoted that, because of Loma Prieta, Bay Area hotels lost about 10 percent of new business. A lot of tourists did not show up, but business travel continued. No study has yet been done, for instance, that relates the loss of tourist trade to the increase in researchers' trade after the Loma Prieta earthquake; there may be a net gain in trade.

Other airport hotels, downtown hotels, and possibly other cities gained (some people may have gone to Los Angeles rather than San Francisco). Hyatt closed the hotel, but they also reduced their costs. They lost revenue, but they also did not have to support the hotel, so their costs were reduced. On the loss side, Hyatt lost profit, and some employees lost their jobs. Vendors who provided services to the hotel, such as food and laundry suppliers, lost. It is also unknown how much they could make up their losses from the other hotels. In addition, there were some additional insurance losses in terms of paying off the employees.

The city of Burlingame lost $120,000 in business tax, but Burlingame gained in building permits for the new work that was done, so that was probably a standoff. Therefore, the question of loss is not easy to evaluate. There are gainers and losers, and under certain circumstances, what happens is not a loss but a *redistribution of resources*; there seems to be a kind of conservation-of-energy principle at work.

Clearly, it is only under certain circumstances that this situation would apply. In this case the loss is really a redistribution, because the general business continues; the people continue to come to San Francisco and the San Francisco airport; the hotel is rebuilt, so that there was a reconstruction project for the construction industry; and the hotel reopened. If the general economy of the region or the country breaks down, obviously this situation does not apply. Also, there are other cases in Loma Prieta where clearly the situation did not apply and where there were significant losses. But, if this sort of individual case is multiplied by 100,000 to 200,000 times, a picture of the complexity of the total economic situation is revealed.

PRESENTATION OF DON G. FRIEDMAN

Three topics will be addressed in this presentation. First, a brief discussion is given of changes in loss-estimation procedures that Travelers Insurance Company has used over the past three decades. There has been significant progress in the development of sophisticated methodologies. Unfortunately, the successful, practical application of these procedures is severely hampered by the lack of appropriate input information, such as geographical inventories of buildings of various types, their damage susceptibilities to earthquakes, and consequent casualty-producing potentialities.

The second topic outlines the need to gain a better general understanding of the major factors that occasionally combine to produce a natural disaster and, subsequently, determine its severity. The need to know more about the disaster-producing mechanism was necessary so that this information could be used as a supplement to, or a replacement for, the often inadequate or inaccurate results obtained from specific applications of the numerical models when appropriate data was not available as input to these computerized procedures.

The last topic is an illustration of the use of natural-disaster knowledge in making risk assessments when sufficient *input* data is not available for the numerical models. This illustration attempts to answer the question of whether a useful estimate currently can be made of the casualty- and damage-producing potentials of low- and medium-rise buildings (insured and uninsured) due to a catastrophic earthquake in the central or eastern United States. It also describes the various types of information that would be needed if a large-scale effort were made to develop a more credible estimate.

Useful recent information includes results of a 1990 Federal Emergency Management Agency (FEMA)-sponsored study on *Estimated Future Earthquake Losses for St. Louis City and County, Missouri*;[1] a 1985 FEMA study on *An Assessment of Damage and Casualties for Six Cities in the Central United States Resulting from Earthquakes in the New Madrid Seismic Zone*;[2] a United States Geological Survey (USGS) 1983 workshop on *The 1886 Charleston, South Carolina, Earthquake and its Implications for Today*;[3] and a 1990 expert group review of a *Metropolitan Boston Earthquake Loss Study*.[4]

To carry out this illustration, there is a need to clarify the meaning of "catastrophic earthquake." Should it be defined in terms of the earthquake's magnitude, its epicenter location, its probability of occurrence, or the casualties and damages that it could produce? In order to examine the use of alternative definitions of a catastrophic earthquake, the fatality and building-damage potentials of a 1990 recurrence of the 1811 New Madrid, 1886 Charleston, and 1755 Cape Ann (near Boston) earthquakes have been estimated, along with a number of lesser-magnitude events with epicenters at the locations of the three major events. The implications of defining a catastrophic earthquake in terms of its physical characteristics such as its magnitude and location rather than the losses that it might produce are examined with the use of a catastrophe index. The use of this index denotes the wide range of uncertainty in loss estimates when sufficient input information is not available.

Uses of Natural-Disaster-Loss Estimations in an Insurance Operation

A large, multiline insurer can have hundreds of millions to billions of dollars of exposure, which is spread haphazardly across hazard-prone areas. Insurance companies make decisions with regard to these risks using whatever data are available. To attempt to answer these questions, a better understanding was needed of the major factors that combine to produce a natural disaster and determine its severity. Numerical modeling and computer simulation techniques have been used to provide this understanding.

In the following discussion, reference will be made to natural disasters caused by hurricanes. Disasters caused by intense hurricanes occur more frequently than high-magnitude earthquakes, but these hurricanes have many similarities with destructive earthquakes, including loss-estimation methodologies.

Loss-estimation procedures for earthquake-caused disasters depend upon the interaction of the geographical pattern of ground motion with the spatial array of the population or properties at risk (elements-at-risk) and their loss vulnerabilities. What happens when an element-at-risk is exposed to ground motion of a given severity is defined as its vulnerability. A hurricane, with its accompanying high-wind pattern, can affect large segments of the population and the built environment. A hurricane-loss-estimation methodology evaluates the interaction of the wind-speed pattern with this geographical distribution of the elements-at-risk and their loss vulnerabilities. Currently, earthquake-hazard evaluations of insured exposures of an insurance company are made using, when possible, ZIP code areas as the geographic designator for locating element-at-risk data. In this way, effects of local influences such as local ground conditions or areas of potential liquefaction can be included in the analysis. However, the elements-at-risk location is not always available within a ZIP code area or even within a county. In these situations, the geographical distribution of the statewide totals must be approximated. How useful is the output of these estimation procedures? It depends on the problem at hand and the accuracy of input information data on the elements-at-risk locations and their other characteristics.

Numerical Modeling of Earthquake-Caused Natural Disasters

In the past, when an estimate was needed of the damage-producing potential of earthquakes for the total building inventory (insured and uninsured) in California and elsewhere, very little credible information could be found on buildings, by type and loss vulnerability, even on a statewide basis. To attempt to answer damage-potential questions on the overall inventory of buildings in spite of the lack of specific element-at-risk information, these losses have been numerically approximated through "what-if" analysis procedures.[5,6]

Early computer simulations modeled the geographical pattern of ground motion on *bedrock* and then superimposed the effects of local ground conditions, which were approximated on a 0.1 degree latitude and longitude grid system, using broad definitions taken from a geology map of California. This approach was encouraging because of the similarities of the simulated ground-motion patterns and the actual isoseismal patterns of past California earthquakes.

The type and quality of available input information for defining an earthquake's physical characteristics, the resulting ground-motion patterns, and the effects of local influences have vastly improved over the past quarter century. The USGS now has computerized estimates of local ground conditions on a much finer scale. In addition, research seismologists and engineers have developed a much better understanding of the earthquake mechanisms and the response characteristics of various types of buildings to a range of possible ground-motion frequencies and durations. The development of new physical measures of ground-motion severity may lead to the replacement of the qualitative modified Mercalli intensity (MMI) scale as a primary measure of ground-motion intensity of future earthquakes. However, for loss-potential evaluations of the recurrence of earlier events, the MMI scale is the only measure that is presently available for estimating ground-motion patterns of these past earthquakes.

The geographical pattern of ground motion of earthquakes can be expressed in terms of physical measures that are specific to various types of buildings--for example, pseudo-acceleration to evaluate damage to low-rise buildings. As a result of these improvements, the numerical modeling and simulation of the disaster-producing mechanism of earthquakes is much more sophisticated than it was in the past. However, a major problem is still the lack of appropriate input information on the elements-at-risk to effectively utilize them.

An additional improvement is in the specification of damage vulnerabilities of various types of buildings to given levels of ground-motion severity. These vulnerabilities can be expressed in terms of structural and nonstructural damage potentials: statistical distributions of the degree of damage expectancy, including the percentage of buildings that might collapse. This latter information is useful in determining the casualty-producing potentialities of these structures. The problem is that these vulnerability characteristics, with a few exceptions,[1,2] have not been determined for building inventories in the earthquake-prone areas, especially in the central and eastern United States.

Another important development in earthquake-loss estimations is the awareness that a moderate-magnitude earthquake can produce the same severity of ground motion as a great earthquake. The difference is that in a moderate-magnitude event, the area affected by this strong motion is much smaller, and the average duration of significant shaking is shorter. The higher-magnitude event is assumed to produce a larger MMI than the moderate-magnitude earthquake, given the same ground-motion severity at a specific location.

Approximating the Natural-Disaster-Producing Mechanism

The second topic deals with attempts to satisfy some information needs by obtaining a better understanding of how and why natural disasters are produced. If one is willing to accept the numerical modeling and computer simulation concept, various what-if questions can be asked. For example, it can be hypothesized that a particular type of building with its characteristic vulnerability has a uniform geographical distribution of maximum possible density. The ground-motion patterns of each of a series of earthquakes of different magnitudes then can be mathematically superimposed upon this building-at-risk pattern, and the overall damage-producing potential of these earthquakes can be simulated. When the estimate of the total loss-producing potential of each earthquake is plotted versus its Richter magnitude, a nonlinear relationship is obtained. This result suggests that a great-magnitude earthquake has a much greater overall damage-producing potential than a moderate- or minimal-magnitude one. Reasons for the nonlinearity include: the size of the area affected, the severity and duration of the strong ground motion, and the mix of ground-motion frequencies.

Fortunately, this overall damage-producing potential of an earthquake, related to its magnitude, is never fully realized, because the elements-at-risk do not have geographical distributions of maximum possible density over large enough areas to be encompassed by the entire ground-motion pattern of the quake. Therefore, the actual realized loss production of an earthquake depends on how the ground-motion pattern happens to overlap the geographical distribution of elements-at-risk. A moderate earthquake centered on the Newport-Inglewood fault in southern California under a large area of densely clustered elements-at-risk can have a much larger actual damage-producing potential than a high-magnitude earthquake on the Garlock fault along the edge of the Mojave Desert or a great earthquake on the San Andreas fault north of San Francisco, where there are fewer elements-at-risk.

The importance of this interaction of the severity pattern of the event with the spatial array of the elements-at-risk in determining actual damage production was highlighted by last year's Hurricane Hugo. If the storm had taken a westward track across the Georgia coastline, where there are relatively fewer exposures, it probably would have been a $1 billion storm. Instead, Hugo directly hit Charleston, causing about $4 billion in insured losses. If it had moved northward across the North Carolina coastline, following the track of highly damaging Hurricane Hazel of 1954, it would have produced about $7 billion in losses. Finally, if Hugo had come up the East Coast with a landfall on western Long Island, with the same intensity that it had at its Charleston landfall, there could have been losses of $18 billion.

The same-strength storm with different tracks had a wide range of possible damage productions, depending upon its final interaction with the geographical distribution and density of the elements-at-risk.[7] Hugo's status as a *catastrophic event* depended on this interaction. The same type of relationship holds for the earthquake hazard. An individual earthquake can have a much different actual damage-producing potential, depending on its magnitude and the location of its epicenter relative to the spatial array and density of vulnerable elements-at-risk. Consequently, the combination of an earthquake's magnitude and its epicentral location relative to the elements-at-risk is of utmost importance in determining its actual loss-producing potential.

This raises the question: what is a catastrophic earthquake or hurricane? Certainly, the physical magnitude of the event is an important factor, but perhaps of equal importance is how its severity pattern (ground motion of an earthquake or high wind of a hurricane) happens to overlay the usually haphazard spatial array and density of the exposed elements-at-risk that are susceptible to loss.[8] A plot of the landfall location of the 247 hurricanes that have crossed the United States coastline since 1870, classified by their physical intensity at landfall as expressed in terms of the five-unit Saffir-Simpson scale, represents a hurricane climatology. If each of these storms recurred in 1990, would all of the Saffir-Simpson code 4 or code 5 storms be considered "catastrophic hurricanes" loss producers? Definitely not! The interaction of severity (high wind) patterns of these storms with coastal clusters of the elements-at-risk determines their "actual" damage production. Many code 3 storms, if they were to recur today, would produce greater losses than the code 4 or code 5 storms because of their particular paths relative to the geographical distribution of the elements-at-risk that are susceptible to damage.

The 1990 loss-producing potentials of all of the 247 landfalling hurricanes, when tabulated against their Saffir-Simpson intensity at landfall, produces a pattern of estimated loss productions of less than $100 million and greater than $1 billion (Table 2-4). There is not a close relationship between hurricane intensity and damage production. For example, 28 percent of the code 3 storms produced simulated losses of less than the $100 million, and 22 percent had a potential of exceeding $1 billion. Because of this analysis, the Saffir-Simpson intensity scale was deemed an inadequate measure of the actual damage-producing potential of hurricanes. Essentially, the scale is an

TABLE 2-4 Percentage of Past Hurricanes with a Simulated 1990 Recurrence that Produce Various Loss Potentials when Grouped by Storm Intensity

	Saffir-Simpson Intensity at Landfall				
Simulated Loss	Code 1 (minimal)	Code 2	Code 3	Code 4	Code 5 (maximal)
< $100 million	82	57	28	5	0
> $1 billion	0	10	22	59	67

indicator of losses to a hypothetically uniform distribution of properties of maximum density.

A "catastrophe index"[9] was developed to provide a more realistic representation of the actual damage-producing potential of individual storms or earthquakes than the physical scales that are currently in use. Table 2-5 lists the computed catastrophe index versus the Saffir-simpson intensity at landfall for the simulated 1990 recurrence of each past storm. A wide range of these damage potential indicies exists within the various hurricane intensity categories. There is a much closer correspondence between the intensity of a landfalling hurricane and its subsequent damage-producing potential, based on a worst case scenario, denoted by a "+" symbol in Table 2-5. Use of the catastrophe index carries the loss-estimation procedure an additional step by taking into account the effects of factors such as the hurricane's landfall location and inland track (or the magnitude and epicenter location of an earthquake) relative to the geographical distribution and damage susceptibilitics of the elements-at-risk.

The geographical distribution of the catastrophe index, assuming a 1990 recurrence of each of the 247 landfalling hurricanes, is very different from the distribution of these storms grouped by their Saffir-Simpson intensity. The catastrophe index analysis also can be used to demonstrate how various combinations of an earthquake's magnitude and epicenter location, relative to the geographical distribution of elements-at-risk, can be utilized to better understand the pertinent characteristics of a catastrophic earthquake in the central or eastern United States.

Loss Potentials of a Catastrophic Earthquake in the Central and Eastern United States

The third topic to be covered is an illustration of some of the earthquake-loss-estimation problems that currently exist because of the lack of appropriate input data to the various loss-estimation models. To begin, an

TABLE 2-5 Catastrophe Index Resulting from the Simulated Present-Day Recurrence of 247 Past Hurricanes (1871-1990) Listed Versus Each Storm's Saffir-Simpson Intensity at Landfall

		Hurricane Intensity at Landfall (Saffir-Simpson Scale)					
Catastrophe Index	Damage potential (current dollars)	Code 1	Code 2	Code 3	Code 4	Code 5	Total
1L	1.00 - 2.49 million	7	0	0	0	0	7
1M	2.50 - 4.99 million	8	3	0	0	0	11
1H	5.00 - 9.99 million	13	2	0	0	0	15
2L	10.00 - 24.90 million	24	10	4	0	0	38
2M	25.00 - 49.90 million	17	10	5	0	0	32
2H	50.00 - 99.90 million	8	11	9	1	0	29
3L	100.00 - 249.9 million	9	14	14	2	0	39
3M	250.00 - 499.9 million	7	5	13	2	0	27
3H	500.00 - 999.9 million	1	2	6	4	1	14
4L	1.00 - 2.49 billion	0	6	10	5	2	23
4M	2.50 - 4.99 billion	0	0	4	5	0	9
4H	5.00 - 9.99 billion	0	0	0	3	0	3
5L	10.00 - 24.99 billion	+	0	0	0	0	0
5M	25.00 - 49.99 billion		+	0	0	0	0
5H	50.00 - 99.99 billion			+	0	0	0
6L	100.00 - 249.9 billion				+	0	0
	Total	94	63	65	22	3	247

The "+" symbol represents the catastrophe index for a worst case scenario in which the insured properties have an unrealistic uniform geological distribution of maximum possible density across the entire area affected by the storm's spatial pattern of high winds.

analysis has been made of the fatality and building (residential and commercial) damage-producing potential of a simulated 1990 recurrence of the 1811 New Madrid earthquake, the 1886 Charleston earthquake, and the 1755 Cape Ann earthquake near Boston. To account for the importance of the combination of an earthquake's magnitude and location relative to the elements-at-risk in determining its actual damage production (i.e., its catastrophic event status), these potentials also have been estimated for a number of quakes of successively lower magnitudes located at the New Madrid, Charleston, and Cape Ann epicenters. A catastrophe index, which has been determined for each of the scenario earthquakes, is used to define the characteristics of a catastrophic earthquake in the central or eastern United States.

Choice of the scenario earthquakes was made by considering the ten strongest events with epicenters in the central and eastern United States during historic times (Table 2-6). Different scales for representing the strength (magnitude) of the earthquakes are listed in the table. In the following discussions involving earthquake magnitude, the scales will be specified, because they are quite different for higher-magnitude events.

Of the seven largest-magnitude earthquakes with United States epicenters, five were located in the New Madrid seismic zone within a 100-mi strip that runs from northeastern Arkansas to southeastern Missouri. Four of these Richter magnitude 8+ events occurred within several months of one another between late 1811 and early 1812. The fifth earthquake, which occurred in 1895, had a lower-magnitude (Richter 6.7) and was located about 30 mi north of the February 1812 epicenter. The other two events had epicenters near Charleston, South Carolina, and Cape Ann, Massachusetts.

A composite map of the maximum ground motion resulting from the four Richter 8+ earthquakes in 1811 and 1812 is shown as the right-hand map in Figure 2-7. It covers a much larger area than the pattern of any one of the individual events that it represents, because the epicenters of the four quakes were not at the same location but were displaced northward along a 60-mi line. As a result, the ground-motion patterns of each event also were displaced northward, thereby overlapping one another. The composite map shows the largest ground-motion severity of the overlapped patterns in each affected locality. It has been assumed, without evidence, that a 1990 recurrence of the 1811 and 1812 seismic activity would be in the form of a single 8+ event and not the series that originally occurred. As a result of this assumption, it was not feasible to use this composite ground-motion map of the 1811 and 1812 events, which was prepared by the USGS in 1985.[10] Note that the USGS attempted to include the probable effects of local ground conditions as indicated by the distorted shape of the ground motion severity pattern in Figure 2-7.

The only immediately available estimate of the ground-motion pattern associated with any of the four 8+ New Madrid events in 1811 and 1812 is the Richter magnitude 8.6 earthquake that occurred at 2 a.m. on December 16, 1811.[11] Because of a lack of observations of the effects of this event to the

TABLE 2-6 Occurrence Date, Location, and Magnitude of the Ten Largest Earthquakes that Affected the Central and Eastern United States and Southern Canada in Historic Times

Date	Epicenter Location	Maximum MMI I_o	Surface Wave Magnitude (Richter Scale) M_s	Body Wave m_b
	United States Epicenters			
Feb. 7, 1812	New Madrid zone	XI-XII	8.7	7.3
Dec. 16, 1811 (2am)	New Madrid zone	XI	8.6	7.2
Jan. 23, 1812	New Madrid zone	X-XI	8.4	7.1
Dec. 16, 1811 (8am)	New Madrid zone	X-XI	8.3	7.0
Aug. 31, 1886	Charleston, SC	X	7.6	6.7
Oct. 31, 1895	Charleston, MO	IX	6.7	6.3
Nov. 18, 1755	Cape Ann (Boston)	VIII	6.0	5.9
	Southeastern Canadian Epicenters			
Feb. 5, 1663	St. Lawrence River region	IX-X	7.2	6.5
Nov. 18, 1929	Grand Banks, Newfoundland	(IX-X)	7.2	6.5
Feb. 28, 1925	St. Lawrence River region	VIII	7.0	6.4

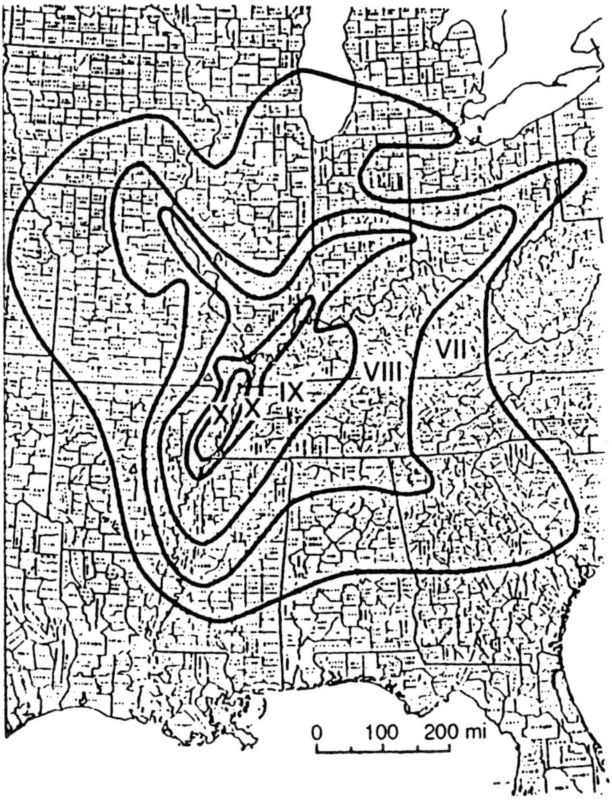

Figure 2-7 Composite map of the highest MMI that might be observed at each location if the magnitude of a simulated earthquake held constant at 8.6 and its epicenter were shifted in increments along the New Madrid seismic zone.

west of the Mississippi River, only a partial ground-motion pattern could be constructed by researchers.

In order to utilize this pattern in the analysis, the ground-motion contours were extrapolated to the west of the Mississippi to provide an approximately symmetrical pattern about the epicenter. The smooth contours indicate that the effects of local ground conditions have not been included. This pattern was superimposed upon a population density map to simulate its overlapping with the current spatial array of the elements-at-risk (population and buildings) as shown in Figure 2-8. Note that the nearest dense cluster of exposures to the strongest ground motion of this particular earthquake is in the Memphis metropolitan area. A Richter 8+ earthquake with an epicenter farther north in the New Madrid seismic zone would generate very strong ground motion closer to the large cluster of exposures in the St. Louis metropolitan area.

Because of the importance of the combination of an earthquake's location and magnitude in determining its loss-producing potential, it was necessary also to estimate the ground motion patterns associated with lesser-magnitude events that have the same epicenter location as the December 16 (2 a.m.) earthquake. As demonstrated in Table 2-6, there is an empirical relationship among the measures of magnitude. The USGS[10] used this relationship to approximate the composite severity patterns of earthquakes of lower magnitude (Richter 7.6 and 6.7) that have the same epicenters as the Richter 8+ event (Figure 2-7). In the conversion procedure, the USGS assumed that the shape of the ground-motion pattern would not change with a reduction in the earthquake's magnitude and that the modified Mercalli intensity could be reduced in the overall pattern by one unit in order to obtain an approximation for a Richter 7.6 event, and reduced by two units overall to denote a Richter 6.7 event.

Similar combinations of earthquake magnitudes and locations were needed to represent earthquake-prone sections of the eastern United States--Charleston, South Carolina, and Boston, Massachusetts. Using the same procedure, the 1886 Charleston earthquake and the 1755 Cape Ann earthquake listed in Table 2-6 were modeled to estimate the loss-producing damage potentials if such events were to recur in 1990. Figures 2-9 and 2-10 show the superposition of these ground-motion patterns on the elements-at-risk spatial array.

Some earthquake experts suggest that a Cape Ann earthquake of magnitude 6.0 is not necessarily the largest possible earthquake in this seismic zone.[3] To estimate the possible loss-producing effects of a higher-magnitude earthquake with a Cape Ann epicenter, the USGS[10] procedure was reversed to estimate the ground-motion pattern of a stronger, Richter magnitude 6.7, quake at the Cape Ann location. One unit of modified Mercalli intensity was added to each of the ground-motion categories defined by Street and LaCroix.[12] The enlarged pattern of the effects of a Richter 6.7 event, superimposed on the 1990 elements-at-risk density map (Figure 2-11) indicates that, although the area affected by the VII-or-above intensities is small

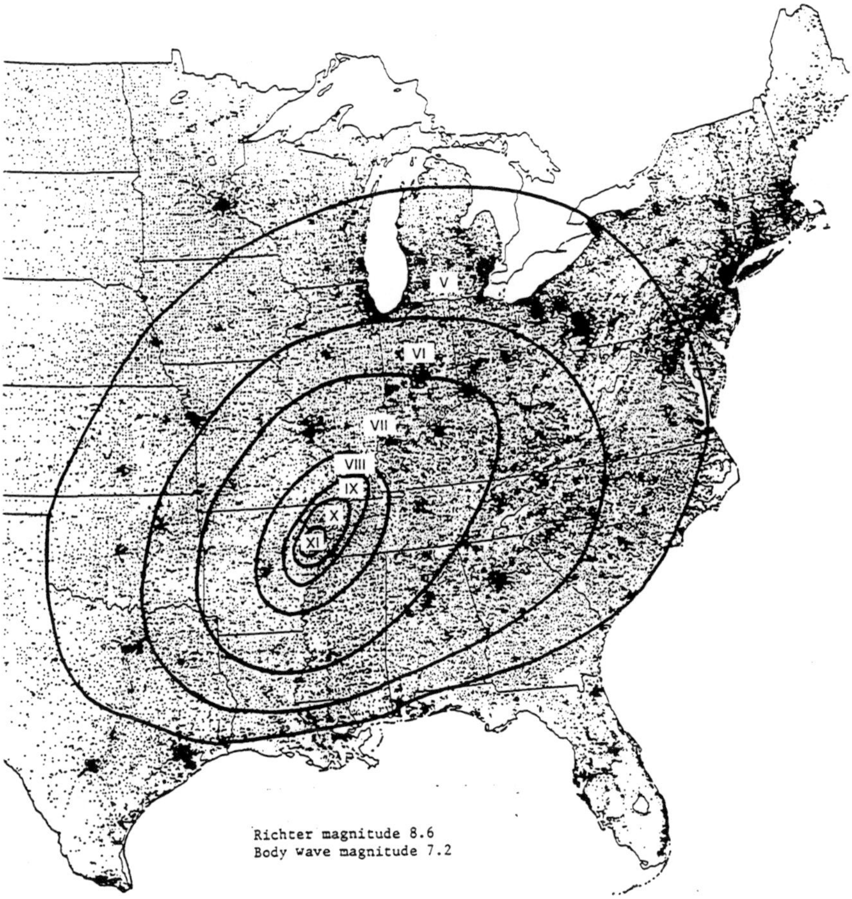

FIGURE 2-8 Loss-producing potential of a recurrence of the December 16, 1811, New Madrid earthquake.

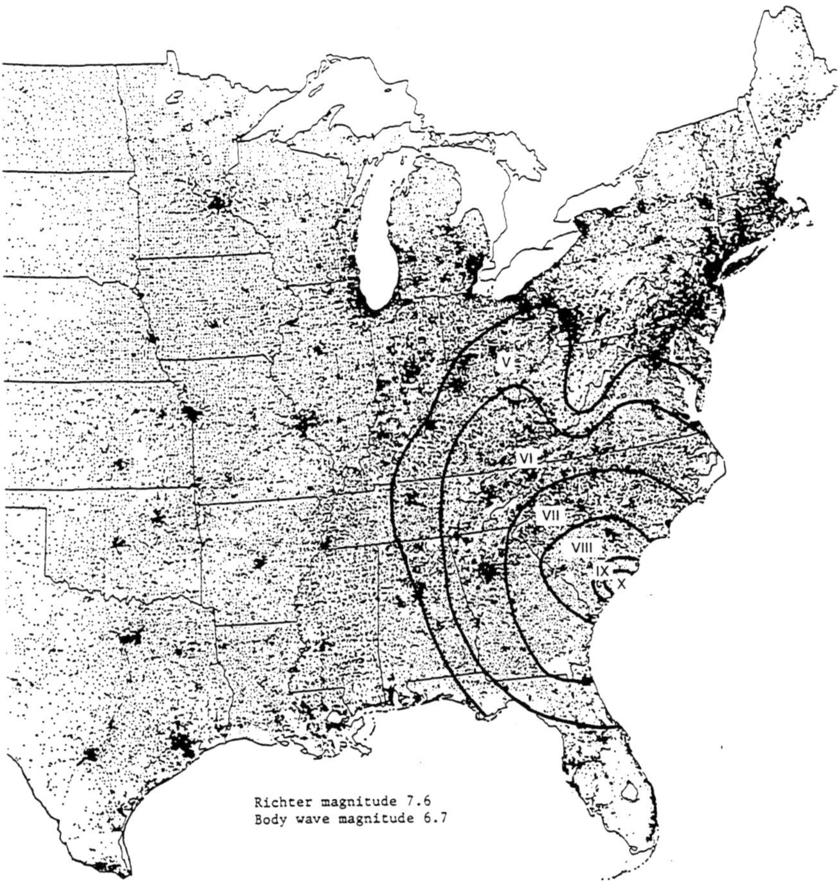

FIGURE 2-9 Loss-producing potential of a recurrence of the 1886 Charleston, South Carolina earthquake.

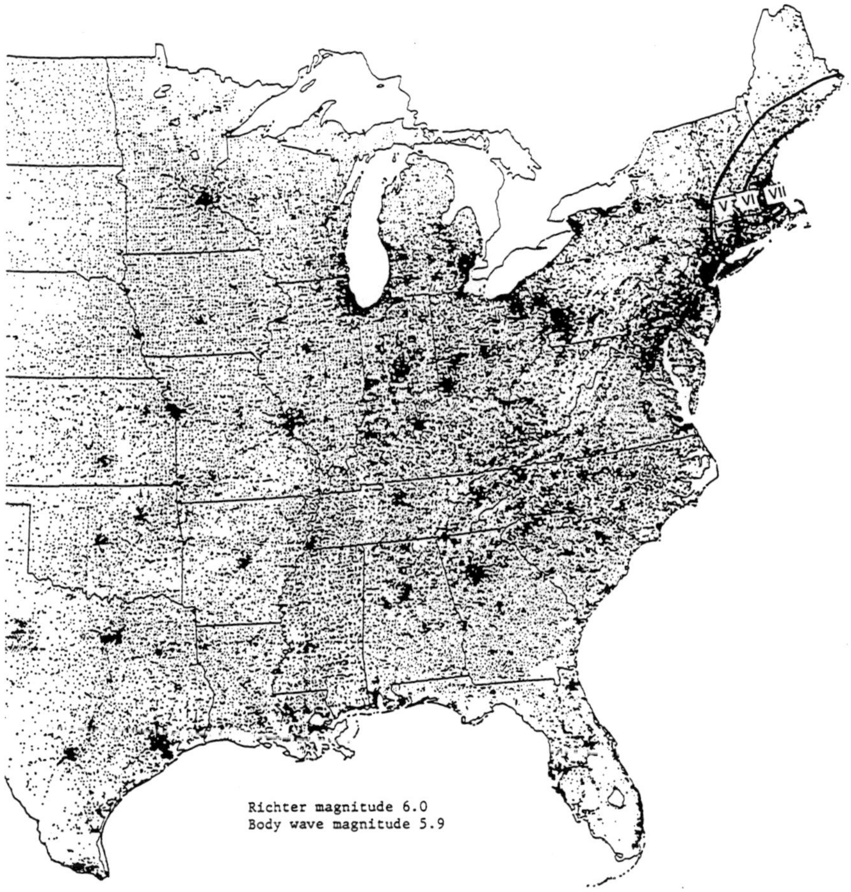

FIGURE 2-10 Loss-producing potential of a recurrence of the 1755 Cape Ann (Boston), Massachusetts earthquake.

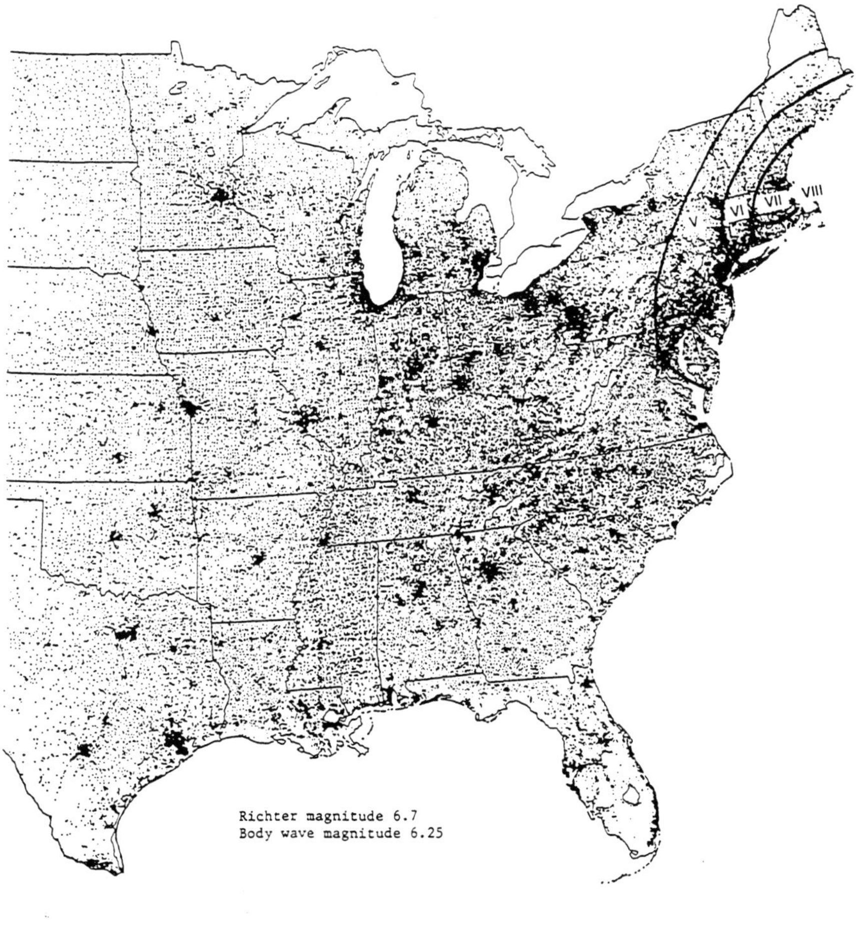

FIGURE 2-11 Loss-producing potential of a recurrence of a stronger (magnitude 6.7) Cape Ann (Boston), Massachusetts earthquake.

compared with that of the New Madrid and Charleston quakes, the strong ground-motion area overlaps one of the largest clusters of densely packed elements-at-risk in the United States.

One of the important information needs for making estimates of losses caused by earthquakes in the central and eastern United States is a mapping of local ground conditions and possible liquefaction areas, as has been done by the USGS in California. Another need, of equal importance, is for the development of a more realistic, physical, measure of ground motion that can be used as a replacement for the qualitative (and in many instances unsatisfactory) modified Mercalli intensity scale. Ideally, this new measure could be translated into an estimate of the ground motion of past earthquakes. However, for the purposes of this illustration, the only ground-motion measures that are available for three of the strongest past earthquakes are the modified Mercalli intensity patterns.

Estimation of Earthquake-Caused Fatalities

Because most earthquake-caused deaths and injuries result from damaged buildings, the casualty-estimation procedure should be based in some manner upon building damage, especially with respect to the percentage of structures that might have serious structural and nonstructural problems. Ideally, an estimation procedure for determining the casualty and damage potentials of these scenario earthquakes should have detailed information on the number and spatial distribution of each type of building in affected areas, along with their characteristics relating to damage and casualty-producing potentialities. Important considerations would be such items as the type and quality of construction, age, condition of upkeep, local ground conditions, building code in effect at time of construction, contents, usage, and number of occupants at various times of the day. Unfortunately, a spatial inventory of buildings and their characteristics is not available in these earthquake-prone areas of the central and eastern United States. However, one of the purposes of this illustration is to attempt to obtain order-of-magnitude estimates of the casualty and damage potentials from scenario earthquakes based on information that is available. The approach taken is described below.

The number of persons within each of the ground-motion categories for four scenario earthquakes (Figures 2-8 through 2-11) plus four others of lesser magnitude were estimated by overlaying in turn the ground-motion patterns on a map of counties in the central and eastern United States. The number of persons within each ground-motion-severity pattern in each of the affected states was summed for each earthquake.

Table 2-7 lists the cumulative number of persons in various ground-motion categories from MMI V-or-more to IX-or-more for each of eight earthquakes. The largest number of persons that would be affected is estimated to exceed 120 million in areas with ground motions of MMI V-or-more during a Richter 8.6 New Madrid earthquake. A Richter 6.0 Cape

TABLE 2-7 Estimate of the Number of Persons Who Would be Exposed to Various Levels of Ground-Motion Severity Caused by Each of the Scenario Earthquakes

	Magnitude		Number of Persons (thousands)				
	Richter	Body Wave	Ground-Motion Severity (Modified Mercalli Intensity Scale)				
Earthquake	M_s	m_b	V or More	VI or More	VII or More	VIII or More	IX or More
New Madrid	6.0	5.9	4,380	2,140	420	130	0
New Madrid	6.7	6.25	21,920	4,380	2,140	420	130
New Madrid	7.6	6.7	50,440	21,920	4,380	2,140	420
New Madrid	8.6	7.2	120,790	50,440	21,920	4,380	2,140
Charleston	6.7	6.25	23,330	11,060	2,830	620	290
Charleston	7.6	6.7	43,530	23,330	11,060	2,830	620
Boston (Cape Ann)	6.0	5.9	17,730	7,190	2,690	0	0
Boston (Cape Ann)	6.7	6.25	45,990	17,730	7,190	2,690	0

Ann earthquake could affect over 2.5 million persons within areas where strong ground motions equalled or exceeded MMI VII. The size of this exposure is much larger than the 400,000 persons that are estimated to be affected by a New Madrid earthquake of Richter magnitude 6.0. If the Cape Ann earthquake had a magnitude of Richter 6.7, over 7 million people would be subjected to ground motion of MMI VII-or-more compared with about 3 million in a Charleston (Richter 6.7) earthquake and about 2 million in a New Madrid (Richter 6.7) event.

Estimation of casualties resulting from each of the ground-motion categories was made by use of fatality rates versus modified Mercalli intensity relationships, which were applied to the number of persons exposed in each of the ground-motion-severity categories within each state for each of the eight scenario earthquakes. The rates were expressed in terms of the number of deaths per 100,000 exposures. Relationships between fatality rate and ground-motion severity were developed from three fatality scenarios:

Fatality Scenario 1: This relationship was developed using an estimate of the number of deaths that might be expected, by state, if the 1886 Charleston earthquake recurred, based on information given in a USGS workshop report.[3] For this illustration, these fatality estimates were related to the assumed ground-motion-severity pattern of the earthquake from which a best-fitting, nonlinear curve was drawn.

Fatality Scenario 2: The scenario 1 relationship based on the South Carolina information was calibrated by use of results of a Boston study, which estimated the number of fatalities that might be expected due to building damage in Boston and some of its suburbs resulting from a present-day recurrence of the 1755 Cape Ann earthquake during working hours on a weekday.[4]

Fatality Scenario 3: This fatality-versus-ground-motion relationship was based on fatality rates that were estimated in FEMA's six-city and St. Louis studies. Estimates of the number of deaths that could occur as a result of building damage caused by a repeat of a New Madrid (Richter 8+) earthquake were made in the 1990 FEMA study of St. Louis city and St. Louis county[1] and in six other Midwest cities in a 1985 FEMA study.[2] Implied fatality rates were determined using the fatality estimates and the estimated numbers of persons in these towns and cities at the time of the simulated earthquake as reported in the FEMA studies.

These rates were then related to the ground-motion severity that was hypothesized by FEMA for each of the locations. Because a range of MMIs was mapped across the study areas, a single, "weighted" MMI was determined for each city by overlaying a grid system on the maps of the seven Midwest cities and St. Louis county. The average MMI for each location was obtained by assuming that the MMI scale is continuous and by weighing various MMI values by the percentages of the total town or city area that they represented. Because of a lack of information on the spatial distribution of buildings within these localities, it was necessary to assume that they were uniformly distributed.

Ideally, fatality rates should be based on expected damage to various types of buildings in areas affected by each of the eight scenario earthquakes. Since this information is not available, it was necessary to assume that the death rates estimated by FEMA in the eight midwestern locations could be related to the attendant MMI and then used as a universal relationship between fatality rate and ground-motion severity for other values of MMI in the central and eastern United States that would be affected by each of the scenario earthquakes. In developing these vulnerability relationships, it was assumed that the fatality rate would be very small when the ground motion was MMI V (1 death per million exposures).

Table 2-8 summarizes state-by-state estimates of fatalities resulting from each of the eight scenario earthquakes using the three vulnerability scenarios. An inspection of this table indicates that there is a multiple of 4 or 5 in the estimated number of fatalities between the lowest values (using scenario 1) and the highest values (using the scenario 3 relationship). A 1990 repeat of the Richter 8.6 New Madrid earthquake could cause somewhere between 7,000 and 27,000 fatalities, depending on the scenario used and assuming that the sets of underlying assumptions are realistic.

Even though the available information cannot provide fatality estimates with a high degree of accuracy, the implied interactions between the earthquake's magnitude, its location relative to the spatial array of the elements-at-risk, and the fatality vulnerability relationships emphasize the importance of considering these particular factors when attempting to define the fatality-producing characteristics of a catastrophic earthquake in the central and eastern United States.

Estimation of Earthquake-caused Building Damage

Estimation of building damage resulting from each of the eight scenario earthquakes also was based solely on the use of immediately available data. Ideally, to estimate building damage due to ground motion, an analysis similar to that carried out in FEMA's six-city and St. Louis studies[1,2] should be done for each city or town in the affected areas. At present, there is a discouraging lack of useful information on various types of buildings, their numbers, spatial distribution, and vulnerability characteristics in the central and eastern United States.

The only immediately available information on the spatial distribution of buildings was obtained by estimating the total value of residential and commercial buildings by county, using data given in a recent report prepared by the Insurance Research Council,[13] which listed the total value of residential and commercial buildings insured against the wind peril in each of the coastal counties along the Gulf and East coasts. This represents a large percentage of the total building inventory. For the purposes of this study, these numbers were related to the population in the counties, permitting the development of a relationship between size of county population and the total value of insured

TABLE 2-8 Estimate of the Number of Fatalities Caused by Each of the Hypothetical Earthquakes and the Three Fatality-Vulnerability Scenarios

Earthquake	Magnitude Richter M_s	Magnitude Body Wave m_b	Number of Fatalities: Fatality-Vulnerability Scenario 1	2	3
New Madrid	6.0	5.9	90	340	480
New Madrid	6.7	6.25	430	1,460	2,200
New Madrid	7.6	6.7	1,990	5,220	7,880
New Madrid	8.6	7.2	6,890	18,110	26,930
Charleston	6.7	6.25	690	2,430	3,620
Charleston	7.6	6.7	3,360	8,490	13,000
Boston (Cape Ann)	6.0	5.9	250	1,260	1,570
Boston (Cape Ann)	6.7	6.25	1,550	5,040	8,170

residential and commercial buildings in the county. The relationship changes with an increase in county population. The spatial extent of the county was not taken into account in the analysis. These relationships for coastal counties were assumed to be universally applicable and were used to convert the population of counties affected by ground-motion patterns of the scenario earthquakes into rough estimates of their residential and commercial building exposures. It was assumed that if county population rather than ZIP code population were used, the mix of the number and type of commercial buildings relative to residential structures as a function of town or city size (urban versus suburban and rural conditions) would be minimized.

Based on this analysis, about 50 percent of the total United States insured residential building values of $6.3 trillion would be affected by ground motions of MMI V-or-more intensity if there were a recurrence of the Richter 8.6 New Madrid earthquake. Strong ground motion (VII-or-greater) would affect about $213 billion of residential buildings caused by a hypothetical Richter 6.7 Cape Ann earthquake as compared with an exposure of $69 billion for a Richter 6.7 Charleston earthquake or $54 billion with a Richter 6.7 event located at the New Madrid epicenter zone. About $6 trillion of the $13 trillion total insured commercial building values in the United States would be affected by ground motions of MMI V-or-more during a repeat of the 1811 (Richter 8.6) earthquake. The overall accuracy of these estimates (based on a conversion from county population to a measure of insured building values) is not known.

Translation of this building-exposure information into a measure of the damage-producing potential of various ground-motion severities was accomplished by constructing and applying three damage-vulnerability relationships similar in form to the ones used for estimating fatalities. These relationships were based on three damage scenarios.

Damage Scenario 1: To obtain an estimate of the minimum damage-producing potential, it was assumed that all of the residential buildings in the central or eastern United States that are affected by the scenario earthquakes would be of frame construction, which has one of the lowest damage susceptibilities to earthquake-caused losses. A relationship given in FEMA's 1990 St. Louis report[1] between frame-building damage expectancy and modified Mercalli ground-motion intensity was used.

Damage Scenario 2: The relationship between ground-motion severity and residential building damage was based on the best-fitting curve through a plot of the implied percentages of value lost for residential-type buildings (insured and uninsured) in the six cities analyzed in the 1985 FEMA study[2] and St. Louis city and St. Louis county in the 1990 FEMA report.[1] The estimated total value of residential buildings in each of these six cities was obtained using the population-versus-insured-residential-building values that were obtained from the coastal county information. These estimates of insured building values were used as an index for approximating the value of all residential buildings (insured plus uninsured as defined in the FEMA studies).

The spatially weighted MMIs for each of the eight areas were used in plotting these implied percentages-of-value-lost data.

Damage Scenario 3: The relationship between residential building damage and modified Mercalli intensity was based solely on information given in FEMA's St. Louis report on expected damage versus ground-motion severity (MMI) weighted by the mixture of residential building types in St. Louis county.

By applying each of these three damage-vulnerability scenarios in the simulated recurrence of the Richter 8.6 New Madrid earthquake, a range of estimated residential damages between $38 billion and $65 billion was found. This range is consistent with an upper bound estimate of about $50 billion for residential building damage made by Algermissen in a 1990 USGS paper.[14] A repeat of the Richter 7.6 Charleston earthquake would cause between $19 billion and $32 billion in residential building (insured and uninsured); between $5 billion and $9 billion of residential damage would occur due to a recurrence of the Richter 6.0 Cape Ann earthquake near Boston.

This same procedure was used to estimate the ground-motion-caused damage potential to commercial buildings. Loss estimates were based on a set of three damage-vulnerability scenarios similar in form to those for residential-type structures. A 1990 recurrence of the Richter 8.6 New Madrid earthquake would cause between $37 billion and $105 billion in commercial building damage. A repeat of the Charleston (Richter 7.6) quake would produce commercial building damages somewhere between $18 billion and $52 billion, and a 1990 recurrence of the Richter 6.0 Cape Ann event would cause between $5 billion and $15 billion in damage.

Estimation of Building Damages by Fire Following an Earthquake

Significant building damages also can be caused by fire that follows some high-magnitude earthquakes. No quantitative estimates or estimating procedures were found in the literature regarding the damage potential of this peril in the central or eastern United States. Therefore, to provide at least an order-of-magnitude estimate for this possibility, an approach used by the Insurance Research Council (formerly the All-Industry Research Advisory Council) in California[15] was adapted for conditions east of the Rockies. It was assumed that the major contribution to fire-caused damage would be from individual buildings or small groups of adjacent structures.

Again, three vulnerability scenarios were constructed based on relationships that were derived from information in the AIRAC study of Los Angeles and San Francisco fire-following-earthquake susceptibilities. In that study, the fire-fighting capabilities in various communities were taken into account (e.g., the number of firetrucks available and the dependability of the water supply under earthquake conditions). Because of the lack of this type of information for the central and eastern United States, it was necessary to assume that a first approximation to the fire-following-earthquake damage production could be obtained by an averaging across the range of individual community fire-fighting capabilities so that the damage-producing potential of this hazard could be related directly to ground-motion severity. It is assumed that, as the duration and severity of the ground motion increases, the number of fire ignitions caused by the earthquake increases, and the capability of the fighters to limit damage to individual structures or the spread of the fire to adjacent buildings decreases, given the need for high-priority search-and-rescue activities for firemen, equipment and communication failures, broken waterlines, and debris and congestion in the streets.

The three fire scenarios were applied to the estimates of total (residential and commercial) building values, by ground-motion-severity categories, for each of eight scenario earthquakes. For purposes of this analysis, it was assumed that residential and commercial building damage caused by ground motion was independent of the fire losses and that the damage threshold is at MMI V, where the average fire-loss potential on nonearthquake days would occur. A repeat of the Richter 8.6 New Madrid earthquake in 1990 would cause fire damages ranging between $7 billion and $25 billion. A recurrence of the Richter 7.6 Charleston event would produce fire losses between $3 billion and $13 billion, and the Richter 6.0 Cape Ann quake could cause fire losses between $300 million and $2.5 billion.

Loss Expectancies Based on Various Combinations of Earthquake Magnitude and Epicenter Locations

Table 2-9 is a state-by-state tabulation of earthquake-caused building damages using the middle (scenario 2) vulnerability relationship for ground motion and fire damage to residential and commercial buildings resulting from a repeat of the Richter 8.6 New Madrid earthquake. Because of the likely low degree of accuracy of these estimates, the relative ranking of the states by damage expectancy is probably more realistic than the absolute values of the loss estimates.

Table 2-10 shows the probability of earthquake occurrence for two magnitudes of earthquakes in the New Madrid zone, the southern United States, and the New England region during the 1990s. Tables 2-11a and 2-11b list fatality and building-damage potentials implied by various combinations of earthquake magnitude and location in the central and eastern United States, using the scenario 2 vulnerability relationships. These potentials increase at different rates as the simulated earthquake's magnitude is increased to its maximum likely value at the New Madrid, Charleston, and Cape Ann epicenters. The rate differences are caused by interactions of the geographical pattern of ground-motion severity and duration with the particular spatial array, density, and vulnerability of the elements-at-risk near each of these three seismic sources.

TABLE 2-9 Estimated Building-Damage Losses by State Resulting from a 1990 Recurrence of the December 16, 1811, New Madrid Earthquake with a Richter Magnitude 8.6, Based on Damage Vulnerability Scenario 2

	Damage (millions of dollars)			
	Fire	Ground-Motion Damage		Total
State	Damage	Residential	Commercial	Damage
Alabama	291	1,152	956	2,399
Arkansas	2,646	8,978	23,726	35,350
Georgia	29	114	112	255
Illinois	853	3,427	3,875	8,155
Indiana	688	2,696	2,318	5,702
Iowa	1	4	4	9
Kansas	1	5	4	10
Kentucky	1,222	4,795	6,082	12,099
Louisiana	206	804	681	1,691
Maryland	0	1	1	2
Michigan	1	3	3	7
Minnesota	0	0	0	0
Mississippi	650	2,683	2,551	5,884
Missouri	1,581	5,854	9,806	17,241
Nebraska	0	0	0	0
New York	0	1	1	2
North Carolina	17	67	67	151
Ohio	462	1,744	1,593	3,799
Oklahoma	38	154	125	317
Pennsylvania	0	2	2	4
South Carolina	9	35	34	78
Tennessee	3,619	13,185	28,055	44,859
Texas	35	138	116	289
Virginia	4	17	16	37
West Virginia	8	30	28	66
Wisconsin	0	1	1	2
Total	12,361	45,890	80,157	138,408

TABLE 2-10 Probability of Earthquake Occurrence in the Decade Before the Year 2001 (in Percentages)

	Earthquake magnitude	
Region	Richter 6.25 Body wave 6.00	Richter 8.25 Body wave 7.00
New Madrid seismic zone	13	2
Southeastern United States	11	2
New England	8	1

The catastrophe index (Table 2-5), which incorporates the effects of these interactions, can be utilized to more clearly denote important characteristics of a catastrophic earthquake in the central or eastern United States. The lowest and highest damage estimates (scenarios 1 and 3) for various combinations of earthquake magnitude and location have been converted to the catastrophe index scale and plotted in Figure 2-12. Hatched areas define the range of estimates for the three epicenter locations, caused by the choice of vulnerability scenario.

If the catastrophe criterion is defined in terms of magnitude of the geophysical occurrence, such as a Richter 8+ event, then the New Madrid seismic zone would be the primary potential producer of catastrophic earthquakes because of its unique capability for generating these great earthquakes. Earthquakes of this magnitude are not likely to occur in either the Charleston or Cape Ann source region, based on current information.

On the other hand, if the catastrophe criterion is based on the relative size of the earthquake's damage-producing potential, then a moderately severe Cape Ann event could qualify as a catastrophic earthquake. Even though its magnitude might be less than that of an earthquake centered either in the Charleston or New Madrid areas, damage production could be greater because of its proximity to large clusters of vulnerable elements-at-risk. Results of the analysis suggest that a moderately severe or high-magnitude earthquake centered at any of these three epicenter locations could cause thousands of fatalities and billions of dollars in building damage.

The purpose of this estimation exercise was to determine the implied casualty- and damage-producing potentials of earthquakes in selected sections of the central and eastern United States as implied from immediately available information. This analysis produced unacceptably wide ranges of estimates because of the lack of pertinent data. It stresses the need for development of the many types of data that are required to efficiently apply the available sophisticated numerical loss-estimation models. The information needs include: the physical characteristics and ground-motion patterns of earthquakes with various combinations of magnitude and epicenter location in the

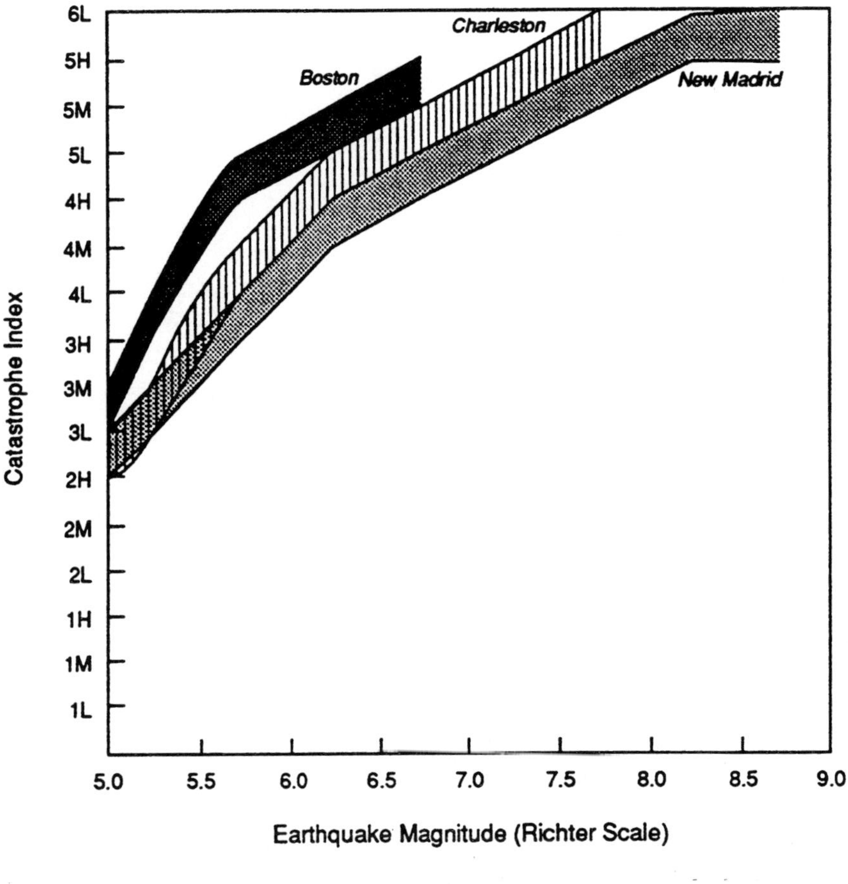

FIGURE 2-12 Estimated damage to buildings caused by ground motion and fire following an earthquake, versus earthquake magnitude. Damage, expressed in terms of the catastrophe index (Table 2-5), is based on vulnerability scenarios 1 and 3.

TABLE 2-11a Estimated 1990 Fatality and Building Damage Potentials in the Central and Eastern United States Resulting from Simulated Earthquakes of Various Magnitudes Centered at the Location of the 1811 New Madrid, 1886 Charleston, and 1755 Cape Ann Events, Based on the Scenario 2 Vulnerability Relationship: Number of Fatalities

Earthquake Magnitude		Epicenter Location		
Surface Wave (Richter) M_s	Body Wave m_b	New Madrid	Charleston	Cape Ann
4.00	4.90	0	0	0
4.50	5.15	0	0	1
5.00	5.40	4	6	13
5.50	5.65	48	70	230
6.00	5.90	340	410	1,300
6.50	6.15	1,000	1,700	3,700
7.00	6.40	2,350	3,800	----
7.50	6.65	4,700	7,600	----
8.00	6.90	9,000	----	----
8.50	7.15	16,500	----	----

central and eastern United States; the effects of local ground conditions on ground-motion severity and its duration; inventories of various types of buildings, their number, and spatial distribution; and their damage susceptibilities to various combinations of ground-wave frequency, severity, and duration.

A final consideration in defining the characteristics of a catastrophic earthquake is probability of occurrence. Nishenko and Ballinger[16] have recently made estimates of the probability of occurrence of major earthquakes in three regions (Table 2-10). Given that an earthquake occurred in one of these broad areas, a conditional probability would have to be applied to determine its chances of being located in one of the three specific epicenter areas discussed above. Nevertheless, based on these estimates, any one of the three source areas is capable of producing an event which could be classed as a catastrophic earthquake in the current decade if loss-producing potential is the criterion.

TABLE 2-11b Estimated 1990 Fatality and Building Damage Potentials in the Central and Eastern United States Resulting from Simulated Earthquakes of Various Magnitudes Centered at the Location of the 1811 New Madrid, 1886 Charleston, and 1755 Cape Ann Events, Based on the Scenario 2 Vulnerability Relationship: Building Damage (Millions of Dollars)

Earthquake Magnitude		Epicenter Location		
Surface Wave (Richter) M_s	Body Wave m_b	New Madrid	Charleston	Cape Ann
4.00	4.90	0	0	0
4.50	5.15	5	5	8
5.00	5.40	65	100	230
5.50	5.65	500	950	3,100
6.00	5.90	2,300	4,300	14,000
6.50	6.15	6,500	12,500	38,000
7.00	6.40	15,000	30,000	----
7.50	6.65	34,000	61,000	----
8.00	6.90	65,000	----	----
8.50	7.15	120,000	----	----

The focus in this presentation has been on fatalities and damages to buildings caused by a high-magnitude earthquake. There are many other sources of loss potential that also have to be considered, such as damage to utilities, roads, and bridges; medical cost for the injured; the cost of debris removal; damages to automobiles and other personal property; business-interruption costs; and liability-loss potentials. Many of these loss potentials are of significant size.[17]

PRESENTATION OF KATHLEEN TIERNEY

Loss-estimation methodologies seek to determine how one or more earthquakes in a given geographic area will affect people, property, and social and economic activity. Approaches vary considerably. Theoretically, the unit of analysis for loss projections can range from the macro- to the micro-level, but most studies done to date focus on communities or slightly larger units. Ground shaking is the most frequently used hazard in these studies, but other primary effects, such as fault rupture and liquefaction, as well as secondary effects, can also be taken into account. Loss-estimation studies can vary in the number of effects they take into consideration; they may, for example, deal only with physical earthquake effects to specific types of structures, or they may be quite broad, taking into account a range of impacts and both direct and indirect earthquake effects. Different strategies for developing inventories of what is at risk and for modeling losses constitute another source of variation in the loss-estimation literature. Nevertheless, although a broad range of potential approaches exists, the loss-estimation studies that have been conducted to date have been relatively limited in the impacts and variables they have considered.

The National Research Council's (NRC) Committee on Earthquake Engineering report, entitled *Estimating Losses from Future Earthquakes*[18] contains a concise review of loss-estimation methods and approaches in engineering and related disciplines that focuses particularly on those methods that are used by local and state governments for hazard mitigation and emergency planning. That report briefly discusses the basic components of loss-estimation studies (seismic-hazard analysis and vulnerability analysis), outlines the elements of deterministic and probabilistic approaches to seismic-hazard analysis, and identifies the types of direct losses and indirect impacts such studies have attempted to quantify. Major categories of losses discussed include building damage, fatalities and injuries, homelessness, and loss to special facilities and lifeline systems. Secondary losses discussed in the report include fire, hazardous materials releases, and indirect economic impacts.

The NRC report and the working papers on which it is based constitute a good overview of available methods and approaches, and they also contain several important discussions on the usefulness and limitations of earthquake-loss estimates. Using these ideas and other insights from the literature as a basis, the following are some general observations about the utility of the work that has been done to date for projecting structural and nonstructural losses, indirect earthquake impacts, and long-term socioeconomic effects, as well as some observations about their relationship to policy. The messages that will probably come across are that existing findings from loss-estimation methodologies have certain inherent limitations as policy tools and that there are currently significant gaps in our understanding of the range of probable earthquake impacts, particularly impacts not related to building damage.

The first topic to consider is how much we know about various types of earthquake losses. There appears to be general agreement among practitioners that loss-assessment methodologies are well developed and validated only for some categories of losses. A taxonomy developed by Petak and Atkisson[19] classifies natural-hazards effects into three categories: (1) primary, such as injuries and damage to buildings; (2) secondary, which follow almost immediately as a consequence of those basic impacts, such as homelessness and disruption of utility services; and (3) "higher-order," which manifest themselves later, such as long-term unemployment and changes in the tax burden associated with disaster recovery. There is considerably more known about impacts in the first category than about those in the other categories. In some subareas, there is virtually nothing known.

There are at least two reasons why this has been the case. The first, of course, is that a considerable amount of loss-estimation research has been done mainly to address the concerns of particular clienteles. The pioneering work in the field[20] was undertaken for the property insurance industry and the government departments responsible for insurance regulation. The field developed in such a way that estimating probable direct damage to property--particularly buildings--thus became a major focus. Ironically, we know considerably less today about the estimation of earthquake-related deaths and injuries than we do about potential building damage. Fortunately, a number of very capable scientists have recently begun efforts to close this gap. When it comes to other social impacts, such as earthquake-caused homelessness, almost no systematic work has been done.[21,22]

Second, while presenting enormous challenges by anyone's standards, estimating direct earthquake effects is in many respects easier than attempting to take longer-term, higher-order impacts into account. When that line of research was developed, building inventories, construction classifications, and other necessary elements for loss assessment already existed, as did a useful historical record on how earthquakes affect buildings. From the standpoint of available data, the picture becomes much murkier as broader impacts are considered. For example, while the task of developing building inventories is difficult, obtaining up-to-date information on building occupancies and uses--which is important for casualty estimation and projections of losses to building contents--is even more difficult. To do this type of work on anything but the most modest scale would require considerably more funding than appears to be currently available.

For some categories of losses, systematic empirical data on events comparable to those of interest do not exist, so there is little in the way of an empirical basis from which to extrapolate. In other cases (e.g., damage to some industrial and to nuclear power plants and defense-related facilities), the data may exist, but access is restricted because of organizational concerns, and the information does not enter the public domain.

Projects such as the major, pioneering loss-estimation work undertaken by the Applied Technology Council[23] attempt to compensate for the lack of a data base by developing estimates based on the judgments of expert panels, with impressive results. Ron Eguchi, a colleague of Professor Tierney's, developed some ingenious ways of getting around data limitations when attempting to model the likelihood and risk to local residents of earthquake-induced hazardous materials releases.[24,25] We are currently attempting to do something roughly comparable for potential earthquake-generated oil pipeline failures in the New Madrid fault zone, with socioeconomic impacts being the main variables of concern. However, efforts like these represent only preliminary steps in assessing secondary losses, and any method is only as good as the data and the assumptions on which it is based. It is also useful to keep in mind that the catastrophic event with which we are concerned here is unprecedented in the United States. A catastrophic earthquake is likely to be qualitatively different from less serious events; the effects generated may be orders-of-magnitude different from those that have served as a basis for experts' projections.

Eminent sociologists who study risk, such as Charles Perrow,[26] have done an excellent job of outlining the pitfalls involved in estimating the effects of accidents and failures--particularly those analyses that focus on infrequent and catastrophic events. Work by people such as Allan Mazur[27] has highlighted problems inherent in trying to base public policy on expert analyses. However, even if disbelief is suspended and a great deal of faith is placed in the methods that have been developed for estimating losses, the problem remains that, for some losses and impacts, there are often simply not enough data from which to extrapolate.

Estimating long-term, regional, or systemwide economic impacts is also complicated by the fact that social systems are so complex. Steinbrugge, in discussing the problem of compiling loss statistics, notes that dollar loss estimates can vary widely, depending on whether losses are considered as "personal" or "impersonal."[20] In other words, the notion of who is likely to bear the costs of damage is interwoven with the cost figures themselves. This is apparent even in the most straightforward cases of physical damage to buildings, and it is even more significant when higher-order effects are considered. In recent U.S. history, the Chrysler Corporation and the savings and loan bailouts show how flexible social systems can be with respect to distributing losses. New public and private initiatives in the earthquake insurance area will likely result in changes in loss projections (and, when the earthquake occurs, changes in how losses are distributed) for various societal sectors. Earthquake investigators may have a reasonably good idea about how a particular earthquake, with a particular intensity, will affect a particular kind of structure and, generally speaking, how much that could cost. Socioeconomic effects are inherently more difficult to model.

Discussions on how to estimate certain sets of economic effects[18] using input-output (I-O) modeling seem to proceed on the implicit assumption that social systems are "closed systems." Even dynamic I-O models are relatively insensitive to changes in the larger environment in which economic subsystems are embedded. Treating societies, regions, and communities as "open systems" could lead to very different loss estimates. In short, without having a clearer idea of what policy options and programs might come about to contain the economic costs of future earthquakes, those attempting to make analyses are at a distinct disadvantage in trying to calculate those costs.

The second topic that needs a closer look is the relationship between loss estimates and public policy. Loss-assessment methodologies appear to be based on the assumption that, since losses can be expressed in terms of dollar figures, persons killed or injured, or other standard scales, the quantitative results of analyses will be immediately relevant for planning and policy purposes. The assumption is that mitigation and planning priorities can be set rationally, on the basis of the numbers the estimates contain. However, the results of quantitative loss estimates will invariably be judged in light of qualitative and value judgments. To use an extreme example, a statistic indicating that 50 persons were killed (or will be killed) due to earthquake damage will be judged very differently by society, depending on whether these are 50 isolated individuals who died from a range of causes, 50 prisoners who died because of the collapse of a correctional facility, or 50 children crushed in a collapsed private school building. To give another, slightly different illustration, the loss of a certain number of housing units of low value--say 9,000 inexpensive units--may constitute a relatively small economic loss for a community in sheer dollar terms, but a very large loss when the community's low-cost housing needs are taken into account. Dollar estimates and raw numbers do not convey a sense of the social meaning of losses.

The literature clearly indicates that risks and losses are not judged in straightforward cost-benefit terms, but rather are assessed by members of society according to a range of criteria. For example, potential losses borne involuntarily are not perceived in the same way as those that are voluntarily assumed. Unfamiliar, uncontrollable, and catastrophic effects are seen as particularly undesirable.[28] If one "loses" $50, that loss will be evaluated in context. It is an entirely different matter whether the money fell out of a pocket, whether it was stolen from a wallet, or whether the $50 was lost because the money was bet on a losing horse at the racetrack.

Damage estimates for geographic areas and categories of assets also tend to mask the fact that losses are typically not distributed evenly in society. Harold Cochrane was among the first to call our attention to what he terms the "distributive" effects of natural disasters. In his paper, Cochrane notes that in disasters, lower-income groups consistently bear a disproportionate share of the losses; they receive, in most instances, the smallest proportion of disaster relief; they are the least likely to be insured (for either health, life, or property); and they live in dwellings which are of the poorest construction and the most subject to damage.[29]

Aggregate statistics on the number of dwellings that will be lost, of persons who will be killed and injured, or of jobs that will be destroyed do not address these disproportionate impacts, which are of great importance when the earthquake problem is viewed from the standpoint of policy. The point here is that similar dollar figures and casualty figures deriving from loss-estimation studies may not in fact be equivalent, in the "social" sense. As an attempt is made to estimate future earthquake losses, and to develop and implement policies to reduce those losses, this is a point that merits emphasis.

A third related issue involves the audiences to which loss estimates are directed and the utility of such reports. Scientists have expended considerable effort in developing and refining loss-estimation methods, but, unfortunately, the outcomes of loss-estimation studies may not be relevant to some potential users, for a variety of reasons. One problem with the existing research is the limited nature of most studies. For example, in attempting to be comprehensive, large-scale loss-estimation studies focus on entire regions. In so doing, they sacrifice applicability to smaller units of analysis and the ability to generate specific estimates. Estimates derived for categories of structures apply only in the aggregate, not to smaller units of analysis. Additionally, budget and data limitations typically constrain the work, so that studies usually focus on shaking only, rather than other seismic hazards, and only on certain categories of outcomes, such as direct losses to buildings. The practical result is that planners and policy makers receive at best only a partial picture of potential losses.

Of course, it can be argued that we have to start somewhere in projecting potential losses, and that any data obtained, however incomplete, is certainly better than none. Nevertheless, policy based on existing estimates is likely to be effective only to the extent that estimates do in fact approximate future losses. What if it is actually the harder-to-measure, poorly understood, secondary earthquake effects that end up actually costing more in the long run? If the very creative loss-estimation work being done by is accurate,[30] earthquake-generated fires are an extremely important area of concern. However, except for his work on San Francisco and Los Angeles, the fire problem has not been studied in depth in the United States. Thus, policy initiatives based on studies of the most obvious, direct effects of earthquakes may not be reducing losses as much as anticipated. At present, perhaps the best that can be done is to be explicit about the limitations of the methods used and about what loss estimates *do not* reveal about overall costs--while at the same time try to put as much emphasis as possible on understanding potential higher-order impacts.

Finally, the Panel on Earthquake Loss Estimation noted several ways in which the scope of loss-estimation studies sometimes limits their usefulness.[18] State officials, understandably, want state or regional estimates, while local officials find most useful those studies that focus on their local areas. Other feedback provided to the panel gave additional insights on why loss-estimation studies have been of limited use. Among the problems cited were: (1) insufficient effort on the part of analysts to involve local officials and policy makers in the loss-estimation process, (2) conflicts and disagreements among experts that undermined their credibility, and (3) the highly technical style in which many reports were written. As has frequently been noted, policy makers and planners are much more comfortable with clearcut decision making criteria than with seemingly vague probabilistic projections. Ironically, the state-of-the-art probabilistic loss-estimation methodologies so favored by professionals may be less "accessible" to users, and less likely to be used, than cruder but more clearcut projections.

Researchers invariably end their papers by talking about the need for more research, and this presentation will not be an exception. It is obvious that, in order to reduce earthquake losses, we need more empirical work to better understand what those losses might be. Presumably, a loss-estimation research agenda will be one of the outcomes of this forum. However, while researchers seek to obtain new data on probable losses, techniques also need to be developed that are sensitive to societal concerns about losses and useful products need to be developed that put different categories of loss into perspective.

PRESENTATION OF ROBERT W. KLING

Types of potential disaster losses can be categorized as follows: (1) direct economic loss, (2) indirect economic impacts, (3) loss of cultural environment, and (4) loss of natural environment. The third and fourth categories are actually components of the first and second categories; however, they are different enough in character that it is useful to consider them separately. This presentation will focus on the third category, loss of the cultural environment.

What is cultural environment? This is a broad notion, but first, a society's cultural environment includes its stock of historic and cultural assets, a broad category of historic monuments, human artifacts, and works of art that are important in providing identity and continuity to a society's culture.

Second, cultural environment includes intangible assets such as human relationship networks and an individual's *sense of place*, which, together, can be called *social capital*. The goal of this presentation is to give some indication of how the value of these kinds of assets can be incorporated into economic analyses of the value at risk from natural hazards like earthquakes.

What is a historic monument or cultural asset? Any definition is likely to be overly broad in some ways and too narrow in others. Yet certainly there is a universal sense of what is meant. As a starting point, think of a house. If it were destroyed, how would the loss be assessed? For most houses, the market value would be fairly easy to estimate, and a good measure of the loss. Alternatively, the replacement cost could be used. Often, these two numbers would give about the same measure.

But what about the most famous house in Fort Collins, Colorado: the Avery House? It is more than 100 years old, and almost everyone would agree that it is worth more than whatever might be implied by its ability simply to supply housing space. What about the White House? Is it worth more to the United States than its value as a luxury residence and an office building? What about a 2,000-year-old hut? It may not look like much, but what is it worth?

Going beyond houses, consider a pottery image of the god Quetzalcoatl, more than 1,000 years old. What is it worth? In the Loma Prieta earthquake, the Museum of Asian Art in San Francisco lost two ancient Chinese vases when they tumbled off their stands. The museum used market-price appraisals to make insurance claims, but what would be done if the Statue of Liberty tumbled off its base?

There are many pictures of the section of the Oakland Bay Bridge that dropped last fall. In that case, no one would assess the loss as anything but damaged transportation infrastructure. But what about the Golden Gate? Is it valuable to society only as transportation infrastructure?

In large measure, the value of such assets lies in their ability either (1) to give a sense of place in history and an identity shared by other people, times, and places, or (2) to provide examples of great aesthetic traditions. These functions are epitomized by the cultural roles played by the Ark of the Covenant or Michelangelo's statue of David. For the purpose of economic valuation, assets like these are differentiated by two key characteristics: they are *irreplaceable* goods, and they are *public* goods. Imagine the importance of this.

A common element among examples of cultural assets is that they are in some sense *unique* and *irreproducible*. Of course, in this case market price observations or other value bases will be few and will be subject to doubt about the effects of elapsed time and transaction irregularities. The surprising prices at which some Impressionist works have sold recently highlight the difficulty. Irreproducibility compounds valuation difficulties. The value of an easily replaced object normally would be near the cost of its replacement production, but what if no replacement can be produced, as for an Egyptian mummy?

A historic or cultural asset is also a public good, even though some are privately owned. Economists traditionally define a public good as one for which consumption is both *nonrivalrous* and *nonexcludable*. Cultural assets typically have both characteristics, and both raise important issues for valuation.

Nonrivalrous or joint consumption occurs when consumption of the object by one person does not at all diminish the object's ability to provide satisfaction to others at the same time. Such is the case with a work of art, for viewing the work does not use it up. Joint consumption affects economic valuation because the value of the object must be measured, not as the value to a single person who would consume it, but as the sum of the values to all persons who would partake in its benefits.

Nonexcludability in consumption occurs when the benefits from a good cannot be limited to those who help pay for it, when the benefits automatically extend to everyone in the community. In the case of a work of art, individuals may be excluded from actual viewing, but certainly not from their cultural stake in the object. In other cases of cultural assets, excludability even from viewing may be impractical.

Nonexcludability affects economic valuation, because it normally prevents public goods from being provided via markets. Therefore, market prices for such goods are seldom observed. The Statue of Liberty, for example, provides a political and cultural symbol for hundreds of millions of people. Though an admission fee may be charged for an actual visit, no private owner of such a monument could ever charge every citizen for the benefits they gain from the monument.

Even when cultural assets such as works of art are privately owned and traded in markets, they retain many public good aspects. Therefore, a market price is unlikely to capture all the asset's value to all the individuals with a stake in art as part of culture.

Thus, cultural assets tend to be unique, irreplaceable, and enjoyed jointly and freely. For all these reasons, market prices are often inappropriate or unavailable as standards of value, and special assessment methods must be employed. For goods traded in normal markets, prices usually reflect either replacement value (i.e., reproduction cost) or benefit value, or both. What about for cultural assets?

Replacement Value

The first candidate as a measure of value is replacement value, or replacement cost. Many cultural assets are irreplaceable assets; consequently, reproduction cost rarely will be an appropriate standard of value for these assets, and one must turn to benefit value.

Benefit Value

While replacement cost relates to the supply price of an asset, benefit value focuses on its utility to demanders. People who act as demanders or beneficiaries of cultural assets will value them for several reasons. These include the benefit they derive from actually enjoying the asset themselves, the benefit from preserving the option to enjoy it in the future, and the benefit they wish to bequeath to future generations. In addition, there may be indirect benefits from the asset's existence, even to those who are not demanders.

Use Value. The most obvious component of benefit value is value in use. This is the dollar value of benefits that accrue directly to people who take the opportunity to enjoy the asset by visiting the site and is the same kind of benefit that normal consumption goods provide their users. Since cultural experiences often are unpriced or are priced on a nonmarket basis, people's willingness to pay for this kind of experience cannot be observed accurately in markets.

Option Value is the dollar value placed on the *potential* consumption services offered by an asset, services that may or may not ever be actualized. People are likely to be willing to pay to have the asset available, in case circumstances lead them to want to see it. Option value is thus a type of conservation value.

Bequest Value is another type of conservation value people attach to assets they are not presently using. In this case, the value comes from the opportunity to pass on potential benefits to future generations. For private assets, willingness to pay for such an opportunity is widely evidenced within families. For irreplaceable public assets, the bequest is necessarily communitywide; individuals' willingness to participate in those bequests is hard to measure, but is important nonetheless.

Tradition Value. In many cases, historic assets and certain other artifacts are instrumental in contributing to the cultural identity of a community or a nation, in ways beyond the apparent value to individuals. This can be called tradition value. The preservation of social continuity and community identity enhance the well-being even of those individuals who would claim no appreciation for the asset based on use, option, or bequest value. In these cases, there is a value in the very existence of the asset, independent of actual or potential personal experience of it.

How can cultural asset values be estimated? Current practices for the evaluation of such goods are few, even nonexistent for many types of assets. For insurance purposes, museums typically value their collections at estimated market prices, yet many museum personnel express the view that some objects are truly priceless, and valuation is pointless. Historic preservation experts indicate that economic value is discussed, if at all, in terms of the economic side-benefits that preservation can bring to a community's economy, and a satisfactory methodology for assessing a historic building's intrinsic value has yet to be developed.

The primary valuation method is market-price appraisals. But since actual prices often are not observed for cultural assets, or since prices often do not reflect the full social value of the asset, substitute methods of valuation are often needed. Some helpful valuation methods adapted from the field of environmental economics include:

- opportunity-cost method;
- contingent-valuation method; and
- travel-cost method.

Market-Price Appraisal

Market price is a natural way to assess value to cultural objects that commonly are bought and sold. In a well-functioning market, the sales price of an object reflects both what a buyer is willing to pay to acquire it and what a seller demands as compensation for giving it up. From either perspective, the price provides a measure of the benefit the asset yields its owner.

Cultural assets are normally unique, making market-price appraisal difficult. However, expert appraisers have established methods of valuation that can usually give reasonably good estimates of market value.

Market prices are thought to be the most objective measure available of how buyers and sellers value goods, and that is an advantage. Though partly subjective, appraisal of an asset's likely market value, based on the actual selling prices of comparable objects, is just one step removed from observation of an actual price for the asset.

An obvious limitation on the use of market-value appraisal is that it cannot be applied to assets for which there are no markets. For example, normal appraisal methods could not yield a market value for the Statue of Liberty. Another important limitation stems from the *public goods* nature of many cultural assets. The price at which an asset sells would understate its full social value. A museum that acts as buyer of art or artifacts on behalf of society may offer a price that reflects social value to some degree, but underfunding of such institutions guarantees that they will be unable to express society's full demand for such objects. The same is probably true in the area of historic preservation. Therefore, for the purpose of social valuation of cultural assets, private-market appraisals can at best serve as lower bounds for the actual values.

Opportunity-Cost Method

A second way to assess willingness to pay for an asset is to tally the opportunity costs of the resources currently dedicated to keeping it. For example, suppose a cathedral in downtown Manhattan is preserved at the sacrifice of $200 million that could be had by converting the land to high-rise office space. Then one can conclude that the landmark is valued implicitly for at least that much by the owner, normally a public entity or nonprofit group acting on behalf of the community.

There are other ways society has shown its willingness to make sacrifices to preserve an asset. The temple of Ramses II, threatened with inundation in the 1960s by the Aswan High Dam, was moved to higher ground at a cost of many millions of dollars, funded by 50 nations and UNESCO. Other assets were left to be submerged, though, apparently a signal that they were not valued so highly. The Acropolis is another monument that is being destroyed by a man-made disaster, air pollution. Apparently the cost of eliminating that threat is too high.

An advantage of the opportunity-cost method is that it uses observable, market-based information, such as land values, to infer a minimum value for a nonmarketed asset. So long as one can have confidence that the owner is rationally preserving the asset on behalf of the public, the inferred value would include option, bequest, and tradition values.

The main problem with the opportunity-cost method is that it often does not measure *policy-relevant* value. The method estimates what people appear willing to sacrifice to preserve the asset; ironically, though, it is measuring the alternative value those people could have if the asset were destroyed. In the case of the cathedral, for example, if an earthquake demolished the structure, the land could be sold for $200 million. To value the cathedral at $200 million would therefore be to imply that the earthquake caused no net loss: a $200 million cathedral is lost but $200 million in commercial real estate is gained. There *is* real net loss, though, which is the amount by which the cathedral is valued beyond the $200 million land cost. Unfortunately, observation of opportunity costs yields no information about the net value of the asset over and above those opportunity costs. The usefulness of the opportunity-cost method is sometimes limited to providing corroboration of values generated by other means. This is not true, however, when using sunk costs, like those involved in moving the Ramses Temple.

To infer the value of asset preservation from opportunity costs, one must have confidence that the owner's decision, whether deliberate or by default, is a rational reflection of public desire. Bureaucratic inertia, for example, might lead to preservation unjustified by real value, and opportunity costs might overvalue the asset. On the other hand, institutional blockages may give signals that undervalue the asset, as in the case of the Acropolis.

Contingent-Valuation Method

A third possible approach is the contingent-valuation method, which is based on direct surveys of a sample of the population (either the "user" population or the population at large). For example, households might be surveyed to discover their willingness to pay to have the Statue of Liberty restored to its original condition and reinforced for its long-term preservation.

In this case, the absence of an observable market value for the asset is handled by creating a hypothetical market in which respondents are asked to make hypothetical economic decisions. The accuracy with which these hypothetical decisions represent real economic valuations depends upon the care with which the survey instrument is constructed.

An important advantage of the contingent-valuation method is its ability to capture the value of the asset to the entire population, including option and bequest values. Furthermore, the measure of benefit given by the contingent-valuation method is a policy-relevant measure; it tells how much people value the asset beyond the opportunity cost they bear to obtain its benefits.

However, because contingent valuation involves surveying a sample of the population and posing hypothetical market scenarios, the method is subject to normal survey biases and to varying interpretations of the meaning of responses. For example, respondents may overstate their willingness to pay for an asset, particularly if they think their statements will affect public policy and yet they may not have to help finance the policy. In any case, the survey instrument must be very thoughtfully constructed and carefully interpreted.

Travel-Cost Method

The travel-cost method is intended as a direct measure of what people are willing to sacrifice to benefit from the asset. For example, one measure of the cultural value of the French Quarter in New Orleans is the amount of time and money people are willing to spend to experience its historical, architectural, and cultural character. This is a variation of the opportunity-cost method. However, the travel-cost method is more useful, for it allows inference of the value people receive over and above what they pay for it; this net benefit is a policy-relevant measure of value.

A more extended example to illustrate this method is a fictional numerical case of a famous old lighthouse. Table 2-12 shows calculation of travel costs, including out-of-pocket expenses and opportunity costs of personal time, depending on how far a visitor comes. Then it relates frequency of visits to that cost.

One measure of the total value of the lighthouse might be suggested by summing the amounts people spend to visit it. In this case, total expenditure (including opportunity costs) amounts to over $1 million per year. One might take this level of expenditure to imply a lower bound for the benefits visitors receive per year, and then infer a value for the lighthouse. But this amount represents the opportunity cost of the lighthouse visits, and if the lighthouse were destroyed, the visitors could receive this same amount of value elsewhere. The real policy-relevant value is the benefits the lighthouse yields over and above the opportunity costs of its services. For instance, Zone A visitors pay an average of $6.30 per visit, though many would be willing to pay more, say $10 or $15.

Table 2-13 uses the relationship between costs and visits shown in Table 2-12, to estimate how much extra visitors would be willing to pay and still come. By estimating how many people would still visit at various cost levels in excess of actual cost, the so-called *consumer surplus* is calculated. In this case, a rough estimate of the annual net benefit to visitors is about $193,000 per year. A final step is to capitalize this yearly benefit, to infer an asset value. For example, at an interest rate of 9 percent, the lighthouse value would be about $2.1 million.

The main advantage of the travel-cost method is that it generates an asset value using concrete, market-based proxies to substitute for the unobserved user price. It is based on actual (not hypothetical) information about user decisions and implied costs. Also, although the travel-cost method is related to the opportunity-cost method, it is superior in its ability to give a policy-relevant measure of the value that would be lost if the asset were destroyed.

TABLE 2-12 Per Trip Costs and Per Capita Visits

Travel Data				Calculation of visit cost per person				Calculation of per capita visits		
Zone of origin	Average round-trip distance from this zone	Round-trip travel hours from this zone	Site visit time	Direct travel cost at $0.15 per mile, 2 persons per car (D)	Opportunity cost of time, at $1.20 per hour (O)	Entry fee (E)	Total visit cost from this zone = D+O+E	Visits from this zone (V)	Population of this zone (P)	Per capita visits from this zone = V÷P
A	20 miles	0.5 hours	1 hour	$1.50	$1.80	$3.00	$ 6.30	6,000	40,000	0.15
B	60 miles	1.5 hours	1 hour	$4.50	$3.00	$3.00	$10.50	80,000	800,000	0.10
C	100 miles	2.5 hours	1 hour	$7.50	$4.20	$3.00	$14.70	10,000	200,000	0.05

TABLE 2-13 Visits Demanded at Various Cost Increments

	At actual cost		If costs were $4.20 higher			If costs were $8.40 higher			If costs were $12.60 higher		
Zone of origin	Cost is:	Visits from this zone are:	Cost would be:	Per capita visits would be:	Visits from this zone would be:	Cost would be:	Per capita visits would be:	Visits from this zone would be:	Cost would be:	Per capita visits would be:	Visits from this zone would be:
A	$ 6.30	6,000	$10.50	0.10	4,000	$14.70	0.05	2,000	$18.90	0	0
B	$10.50	80,000	$14.70	0.05	40,000	$18.90	0	0	$22.40	0	0
C	$14.70	10,000	$18.20	0	0	$22.40	0	0	$26.60	0	0
Totals		96,000			44,000			2,000			0

The travel-cost method has several disadvantages and limitations. Because the value estimate is based on information only about actual users of the asset, it indicates only value in use, and not option, bequest, or tradition values. Those who value the option of visiting a site, but have not yet done so, have made no observed choices that will be incorporated into the data. Second, the method is difficult to apply to an asset that typically is visited in conjunction with several others during a single trip, because the total costs of the trip must somehow be allocated among the various sites visited. This complication would occur in applying the method to value each of the monuments in Washington, D.C., or each of the paintings in an art museum. For these reasons, the travel-cost approach would be used most easily when visiting the asset site requires a short trip dedicated to that asset alone, or when one wishes to value a group of assets collectively.

This discussion has been intended to give a flavor of the methods that might be considered to assess cultural asset values. None of these approaches is yet very refined. Analysts hope to explore their potential in the near future.

Loss of Social Capital

The term *social capital* has been used many different ways. For the purpose of this presentation, social capital is the set of human ties that a community's members find in that place. These ties can be of at least three types: friendships, professional relationships, and an internal sense of stability or home. Such relationships might at first be considered entirely noneconomic and beyond the scope of a tallying of economic damages. However, their economic value can indeed be assessed, because this social capital is an extremely valuable asset that people evidence a very high willingness to pay to preserve.

A piece of capital is an asset that yields income. We think of social ties as capital in the sense that they are assets that yield psychic income that would be important to a household along with pecuniary income. Now, psychic income is a concept rather well developed by economists, and it can have many sources other than social ties: job conditions, public amenities in a place, natural beauty, climate, etc. But this psychic income from social capital is different in at least one key way; it is nontransferable, since it is attached to specific individuals and is developed historically over time. If one family moves out of a community and another just like it moves in, the new family could take over enjoyment of the mountains, the town, the schools, etc. just as much. But they could not take over the human ties.

The destruction of social capital could be a natural consequence of the forced migration that could follow a major disaster; one might think of such capital as a special type of irreplaceable cultural asset, but one that is more personal and intangible than those discussed earlier.

The evidence of social capital in our society is widespread. The best evidence is the fact that people like to stay where they are, despite many incentives to move elsewhere. There are all manner of natural disasters, as well as economic disasters like plant closings and agricultural depressions, that one might think would be a large inducement to relocate, . . . but people stay.

It is impossible to deny the existence of social capital. But what is the importance of this social capital from a public policy perspective? When there is a disruption, when a move is forced, a valuable asset is destroyed. Prevention of that destruction then has a value that might be included in the benefits side of various benefit-cost-based public policy decisions, just as the protection of material assets is included. This is an element that is rarely explicitly included in formal benefit-cost analyses.

In the area of natural-hazard management, we can see that hazard-mitigating types of investment decisions might be made to take into account the value of saving a town, where the town is viewed as more than just a collection of real estate and physical capital. Even programs normally denounced, for imposing the condition that the relief be applied to rebuilding on the original spot, may be seen as serving (perhaps to excess) the objective of preserving social capital.

Social capital is not bought or sold. So, as for the assets mentioned earlier, less-direct methods of assessing its value will be required. One might use a contingent-valuation approach, surveying people to find out what their social capital is worth. Alternatively, one might look for ways social capital affects economic decisions.

Two markets that will be affected by migration are housing markets and labor markets. Residents' hesitance to leave may be reflected in the prices at which they are willing to sell their home.s Or, their desire to stay may show up in the wages they are willing to accept; it might take surprisingly large wage drops to induce people to leave their hometowns. In economists' terms, labor supply would respond inelastically to wage drops. This idea can be expanded upon as an example of how an economist might use objective market data to estimate intangible, subjective values. These ideas probably make more or less sense, depending on whether one is more or less familiar with the economists' concept of supply and demand curves, and their interpretation.

The phenomenon common to the three examples cited earlier is the fact that many communities appear to face a dual labor-supply curve, or what might be seen as a kinked curve. In growing, these communities find that a small wage differential relative to other regions will induce a sizable labor influx (Figure 2-13). Labor demand shifts out, wages rise a little, and the labor force grows a lot.

But a reversal of economic growth generally will not lead to a labor force shrinkage that is so elastic (Figure 2-14). If labor demand drops back down, wages and employment will move, not back along the original labor supply curve, but along the steeper, dashed curve. The result is that wages drop more significantly, but emigration is not significant. The clearest explanation for this dichotomy involves the stock of social capital that a region's inhabitants establish as they spend time there. Reluctant to abandon this social capital, which has a significant human value, workers are willing to accept more-

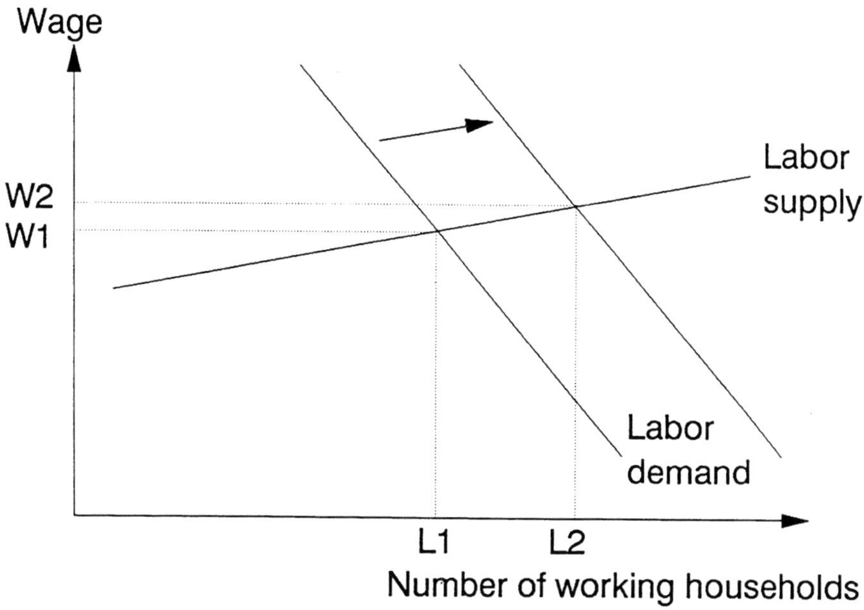

FIGURE 2-13 Effect of a local labor demand increase.

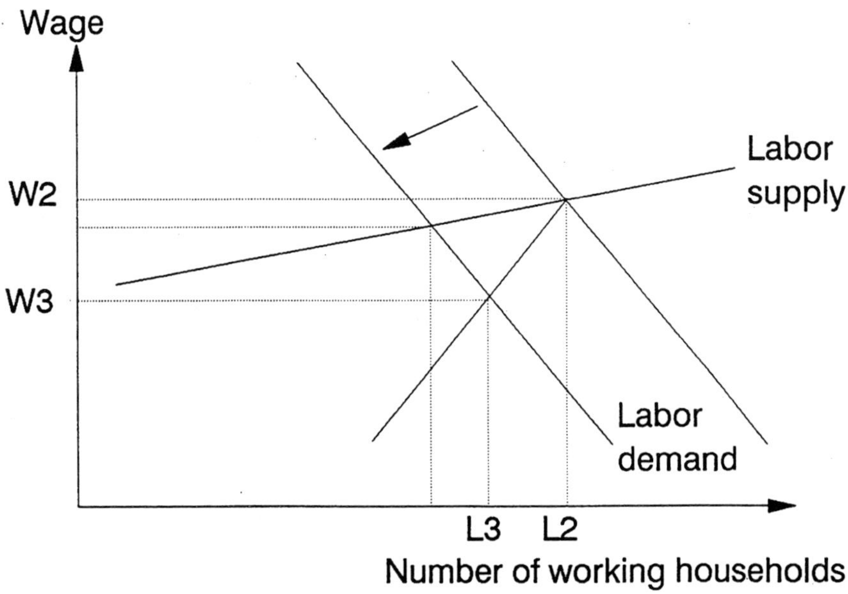

FIGURE 2-14 Effect of a local labor demand decrease.

severe wage cuts in order to remain in what has become their home community.

Figure 2-15 is a more elaborate version of Figure 2-14. For simplicity, the labor-supply curve is drawn horizontally. Labor demand is shown as falling by steps, with more workers leaving the labor force at each step. In each case, the heavy line segments indicate the wage gap that is required to induce that marginal worker to move somewhere else. It thus measures the money the worker is willing to accept elsewhere in compensation for giving up his or her ties locally. If labor demand falls all the way to the lowest level (L3), and if the social capital value is added up for all the workers who leave, the value of social capital lost would be the area of the shaded triangle in Figure 2-16, which could be measured if the labor supply and demand curves were known. This is a loss to those who were caused to move.

The shaded triangle, by the way, measures the wages lost by those who still choose to stay. That is an indirect economic loss of the sort that will be discussed in Chapter 3.

It is important to note that a disaster need not cause lower labor demand for this analysis to be relevant. Even if emigration were caused by destruction of housing stock, for example, estimation of labor supply and demand curves could indicate the amount of social capital at stake.

In measuring social capital in this way, adjustments would have to be made for at least two important factors. First, labor immobility could be due to other ties that are not social capital: equity in a house, for example. That would have to be accounted for. Second, some communities might be better able to afford wage drops, to preserve social capital, than others. By affecting how much they show a willingness to pay for social capital, that would affect the implied dollar value of that intangible asset, in a way that might not be justified on philosophical grounds. Economic values might be adjusted accordingly.

In conclusion, there are important intangible benefits that are at risk from natural hazards. Though they are intangible, there are promising methods available for assessing their value in economic terms, so that they can be included in economics-based policy analysis.

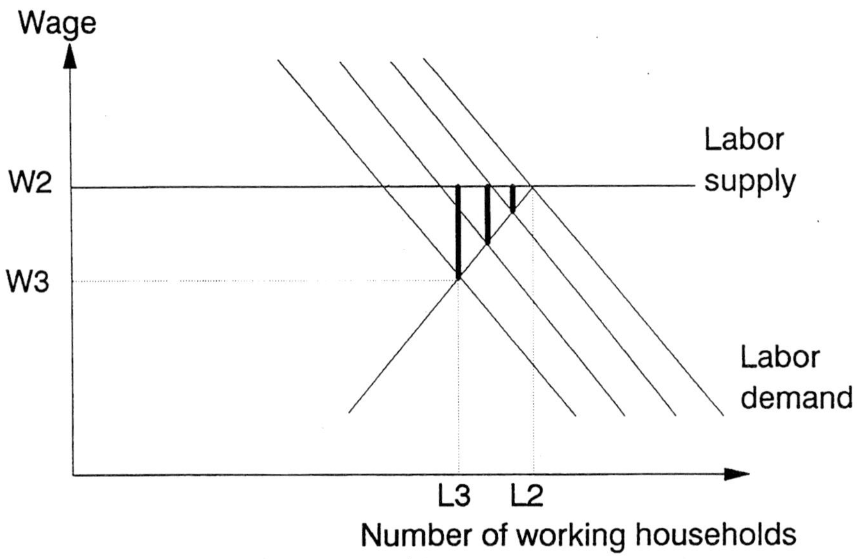

FIGURE 2-15 Social capital lost from relocation.

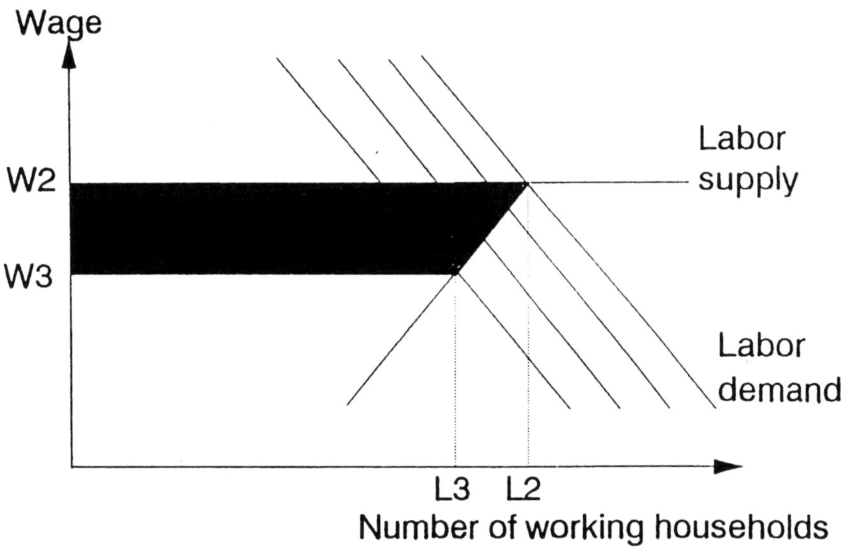

FIGURE 2-16 Losses to workers from lower labor demand.

GENERAL DISCUSSION OF CHAPTER 2

QUESTION: A number of the issues you raised are quite rich, and all of them could be discussed at great length, but I have just a couple of points. First, some of these issues have been discussed at great length by NAPAP, The National Acid Precipitation Assessment Program. For example, Joel Shirago of EPA did a fairly careful analysis of some of these issues and the approaches. I think we have a long way to go before they come up with really useable results, but there will be an economic assessment done on the impact.

Second, it is amusing that the Cape Hatteras Lighthouse was referred to, because the Academy undertook a major study for the National Park Service of the cost associated with moving the Cape Hatteras Lighthouse or protecting it with copper gowns and so on. Many of the disaster scenarios on the erosion effect will go into the next study.

DR. KLING: That is interesting because the National Park Service has a duty to assess historic assets; yet their reports indicate that they are full of great assessments of the cost of protecting or restoring these assets, but there is nothing about benefit. Is it worth a million dollars? Is it worth $20 million? Is it worth $100 million?

QUESTION: This is for Don Friedman. Does Travelers Insurance currently use your models to determine insurance rates or what markets to participate in? In other words, are these models now influencing the actual insurance policy of the company?

DR. FRIEDMAN: For the last 35 years I have attempted to answer questions that management has on the various effects of natural disasters. This is one input that would go into whatever other considerations they might have in terms of competition or what have you. Yes, this is an input into the decision-making process.

QUESTION: Dr. Friedman, have you tried to test your method against the results of the Loma Prieta earthquake?

DR. FRIEDMAN: Yes. Simulation modeling is never finished because it is a continuing process. Each time you get a new event, you try to look at it in terms of verification, calibration, and so forth. Fortunately, it worked out well.

QUESTION: Maybe I should change my question. When you did that, how big a change did the Loma Prieta have on your model? I was asking whether your procedure would have given a similar loss estimate to what the actual losses were from the Loma Prieta earthquake? In other words, how good is your model?

DR. FRIEDMAN: Good is a relative thing. From the point of view of decision-making, we were very satisfied with the results. The insurance industry cannot afford to go to each structure and do an engineering analysis. What is going on below the surface is still unknown. But from the point of view of an insurance operation, the law of large numbers rules.

QUESTION: This is addressed to all four panel members from your various points of view. Each of you, in some way, addressed the secondary effects that occur from earthquakes. These effects are obviously very difficult to insure against. They are hard to measure, etc., and yet we as a society have to deal with them in some fashion. I wondered if there were some thoughts that the panel had about how we address these effects in terms of both the concept of loss mitigation and also with the question of the role of government and who should pay, who should be involved in this? I think these are questions we are going to have to address. All the remarks were very stimulating in terms of raising those issues.

DR. FRIEDMAN: One thing I did not get a chance to discuss was the payoff by various lines of insurance for a large insurance company. How many assets would a company have to sell off in order to settle, perhaps, hundreds of millions of dollars in claims? You would look at the various distributions in terms of types of insurance. This reduces the ability of these companies to sell more insurance because of the ratio between the amount of money they have on reserve and the amount of premiums they can put out into the field. There are obvious difficulties that must be implied from a catastrophic event--a great earthquake, a great hurricane--in terms of the ability of individual insurance companies to continue to sell insurance in other parts of the United States in addition to the area where the event occurred.

DR. ARNOLD: There is a balance sheet. The balance may not come out exactly, but one has to look at the balance sheet. I know there is going to be a lot of focus on losses. However, there are also gains. The insurance company tends to operate as a closed system and look only at its own balance sheet.

If you look at the whole universe, it is an open system, and we must look at the gains, even though they do not balance out. In terms of Dr. Kling's presentation, it is great to see an economist using architectural arguments; but even there, there are gains. For instance, I would argue that Mexico City is a much better city environmentally now because of the earthquake. There are two reasons: (1) it now has a lot of small parks which are environmentally very satisfactory in a city which was very short of open space; and (2) it now has 44,000 units of well-designed, affordable housing which it did not have before. Both of these things were forced by an unfortunate circumstance, but if you look back at the history of cities and the evolution of cities, you will see that earthquakes and catastrophes are part of the natural process. So, we should keep an eye on the balance sheet as we proceed.

DR. TIERNEY: We are continually hearing about how expensive it is and how difficult it is to study certain kinds of problems, to develop certain types of inventories, to look at certain kinds of variables. Wouldn't it be nice if there could be some modest efforts directed at trying to get a handle on some of the more elusive effects, just to balance off the tremendous emphasis on some other types of earthquake-damage effects that we see. This is a plea for more empirical research in this area.

3

Overview of Economic Research on Earthquake Consequences

This session is devoted to a state-of-the-art overview of economic research efforts that have focused on the consequences of major earthquakes. The purpose is to identify what is known about these consequences, both theoretically and empirically. What are the strengths and weaknesses of these efforts to date, and what gaps remain to be filled?

Professor Harold Cochrane, a professor in the Department of Economics at Colorado State University, will present this overview. Dr. Cochrane has extensive research and advisory experience in environmental issues and natural disasters, especially with respect to the distributive effects and economic consequences of such events. He has been involved in three recent NSF-sponsored projects investigating a variety of economic issues on natural-disaster losses.

PRESENTATION OF HAROLD COCHRANE

The state of the art of economic research on the consequences of a catastrophic earthquake is both abysmal and elegant, the former because data are seldom collected in a form consistent with accepted loss-accounting principles, and the latter for the level of mathematical sophistication attempted. This overview will be devoted to the often forgotten but essential principles for assessing impacts and will avoid technicalities, serving instead to highlight a broad range of issues which go well beyond those included in conventional damage assessments.

The issues presented are the product of several NSF-sponsored projects to develop a guidebook for practitioners conducting damage assessments. The book is entitled, *Damage Handbook: A Uniform Framework and Measurement Guidelines for Damages from Natural and Related Man-made Hazards*. The need for such a handbook was identified by Eleonora Sabadell. Its application to the Loma Prieta earthquake was made possible by William Anderson at the Foundation.

Earthquakes produce physical consequences. Tens of thousands of people may be affected, either directly or indirectly. Secondary events such as fire, landslides, and dam failures may widen the scope of damage, possibly including injury to fragile ecosystems. However, from the economist's viewpoint, the mere description of physical consequences is just the beginning of a damage assessment. The most challenging task, as Professor Kling so ably presented in Chapter 2, is that of valuation. This overview will not tackle the valuation problem in any depth, since it would only repeat many of the points raised in Professor Kling's presentation. It will focus, instead, on a set of principles, distilled from an integrated accounting framework. Each will be discussed in turn and will be applied to the Loma Prieta earthquake. Finally, the impact of the Great Tokyo Earthquake of 1923 on the U.S. economy will be speculated upon.

Useful Concepts for Identifying and Measuring Economic Damages

Catastrophic earthquakes typically cultivate images of death and devastation, which might be extrapolated to include a litany of economic impacts. Although each clearly reflects a different aspect of the event, they are not additive. Double-counting and miscounting losses is a problem so severe that little credence can be placed on current damage estimates.

The simple relationships of accounting identities and budget constraints, which govern a regional economy, are the fundamental building blocks from which economic damages are isolated and measured.

Income Accounting and Damage Assessment

In its simplest form, a regional economy consists of a producing sector, consumers, and government. The producers employ members of the household to work with the existing plant and equipment. Incomes are paid to the laborers and those who have capitalized the firm (i.e., the firm's financiers) are entitled to wages, interest, and dividends. The demand for products provides the firm's owners with the incentive to continue producing, which also means that the workforce, plant, and equipment will continue to be utilized.

Not surprisingly, the demand for products stems from the incomes earned. The presence of government causes some spendable income to be diverted from households, but as Figure 3-1 shows, government purchases and payments produce a new set of demands, which may or may not be equivalent to what households would have done in the absence of government.

Without doubt, this is the simplest characterization of economic activity. However, even at this abstract level, several important principles are worth noting. First, the level of economic activity can be measured by counting expenditures, or incomes, but not both. Income to the firm's owners, its workforce, and the firm's financiers must be equivalent to value of the products produced. This is because the price of a product reflects all the costs incurred in its creation, which in this case is the sum of wages, interest, and profits. This simple result provides an important loss-accounting guide: damage assessments could focus on incomes lost or spending lost, but not both. Either should yield the same result.

This schematic of the economy also explains why lost sales taxes and the local government services such taxes support should not be added to lost income. Double-counting is involved here as well, albeit in a more subtle form.

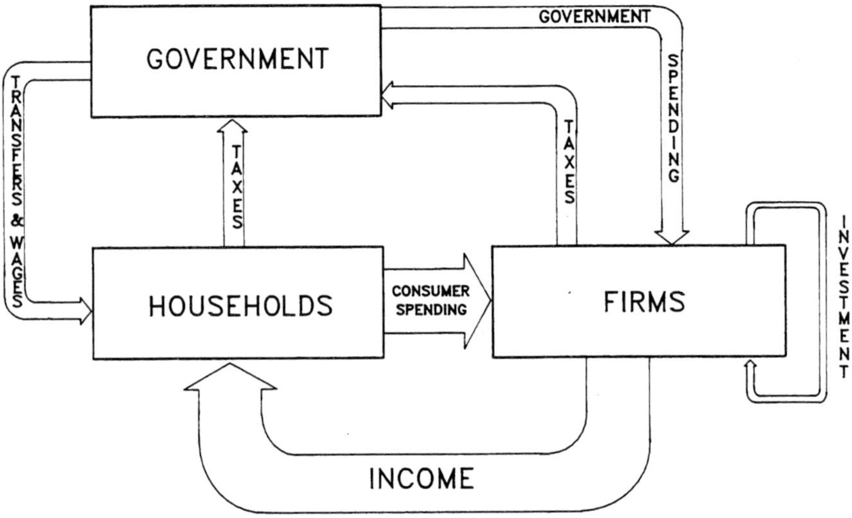

FIGURE 3-1 Flow of payments in a simple, three-sector economy.

Since the primary source of government revenue is traceable to income (sales and income taxes) and property value (property taxes), a disaster-induced reduction in these values implies a proportionate reduction in government revenue, with one minor exception, the latter is derived from the former. It is therefore incorrect to add loss of government revenue (and the associated services) to property and income losses. To do so would count losses twice.

This description is, of course, oversimplified, since regional economies are embedded in a much broader and more complicated matrix of economic forces. For example, households may attempt to spread their losses over time by liquidating real or financial assets or going deeply into debt. Insurance payments, philanthropy, and federal aid will tend to soften regional losses, thereby masking the real effects of a disaster. These possibilities illustrate the pitfalls inherent in loss measurement and further underscore the need for a carefully developed framework.

Direct Versus Indirect Economic Impacts

The so-called indirect effects of disaster are often confused with the direct impacts. Direct impact refers to the loss of plant and equipment which stems directly from damages sustained in the event plus any associated loss of employment. These losses may produce supply bottlenecks, which may produce a ripple effect, inducing layoffs in related but undamaged industries. Such effects have not been widely observed, although some have speculated that the loss of semiconductor production in Santa Clara would produce such effects. Even this scenario seems farfetched today, given the recent emergence of Japan as a major world supplier. It may be that the Great Santa Clara Earthquake simply eliminates the United States as a chief force in the computer chip industry, a trend which has been clearly evident over the past decade.

Reductions in household incomes resulting from layoffs, bankruptcies, and bad loans would produce a separate set of economic effects, referred to as induced or multiplier effects. These too are lumped under the category indirect effects. And, these too have been difficult to detect in Loma Prieta and Hurricane Hugo. The decline in tourism suffered by the barrier-island resorts of Isle of Palms and Sullivans and Folly islands was more than compensated by the huge injection of insurance claim money into the state. By all accounts, South Carolina is experiencing an economic boom of sizable proportions. The same cannot be said for the Bay Area region, since the amount of outside resources, insurance, or federal aid funnelled to the region has been comparatively meager. Even so, the secondary economic effects of the earthquake are nearly impossible to detect.

There are two reasons why secondary losses are so difficult to detect: (1) the economy is more resilient than most economists like to believe, and (2) the effects of disaster are shifted to other regions or to another time period (possibly to other generations). In a world of federal budget limitations, increased aid to help earthquake victims means either tax increases for the general population or a reduction in spending somewhere else in the nation. Even if lawmakers can successfully negotiate an increase in the budget deficit by keeping the aid off the budget, as they did with the savings and loan bailout, the nation is still left with the tax liability which the debt represents. In this instance, the costs of reconstructing an earthquake-torn city are left to our children. Multiplier effects and related secondary losses are, therefore, either pushed onto other regions or shifted in time. They are not eliminated.

Real Versus Financial Effects

Many have speculated about the real effects of disaster, but these effects have been seldom observed. There are several instances where a natural disaster has triggered a depression. Two of these instances are: (1) the Dust Bowl period of the 1930s in the United States, and (2) the Managua earthquake. But even in these instances, the natural event was accompanied by man-made events, the stock market crash and a political revolution, respectively. Counter examples are clearly more plentiful.

So, why do some disasters trigger or at least accompany severe economic contractions and others do not? The answer lies in the underlying strength of the economies affected. The United States has learned from Managua, Nicaragua, and Johnstown, Pennsylvania, where it has been observed that disasters tend to accelerate ongoing economic and social processes that were working prior to the event. Failing economies experience a sudden collapse, whereas robust economies experience a boom. There are sound economic reasons for these observations which will be addressed later in this presentation.

The financial effects of disaster are simply repercussions of the disaster on the stock and bond markets. There are many reasons why stock and bond prices move, but clearly psychology or the state of expectations plays an important role. Ignoring this important issue temporarily, the occurrence of a large disaster may or may not impact these markets. Clearly, the value of corporations directly impacted by the disaster will suffer, for they have lost productive capital. And, to the extent that the market value of the corporations affected reflects the valuation of income streams these companies are capable of generating, their stock prices should decline. In a rational world, these markets would decline by the value of the capital lost.

The total loss from Hurricane Hugo and the Loma Prieta earthquake combined amounted to under $10 billion, nine-tenths of which was sustained by homeowners. The value of the equity market in this country is over $2 trillion. This means that these disasters may have produced a .005 percent change in the capitalized value of corporate America, an insignificant amount to register on the stock exchange. This is not to say that some corporations, such as insurance carriers and resort companies in South Carolina, were not impacted. But overall, the effects were negligible, particularly in contrast to the wild swings in the New York and Tokyo stock exchanges, which caused nearly $2 trillion in paper losses combined.

Loss-Measurement Principles

The following seven loss-measurement principles reflect the conceptual orientation just presented.

1. Measure Loss With and Without the Earthquake, not Before and After. This is one of the clearest and simplest economic principles guiding impact assessments. It is also one which is commonly violated, however. Loss studies are, of necessity, conducted over long time intervals, during which economic pressures unrelated to the disaster can mount. Because of this, it is possible to conclude that a disaster produced an economic change, which more rightly should be attributed to unrelated but correlated factors. U.S. Steel's decision to shut down the Johnstown mill after a devastating flood in 1965 was linked to a chronic decline in the plant's profitability, rather than the sudden onset of damage. The event simply provided management the excuse to terminate operations, a decision which would have eventually been made with or without the disaster. Under such circumstances, it would be incorrect to attribute loss of jobs and the accompanying economic downturn to the flood.

2. Do Not Double-Count Impacts. The accounting framework discussed earlier pointed out several ways in which losses could be double-counted:

- *damage to assets versus flows*--The destruction of productive capital reduces the region's flow of income. Both cannot be counted.
- *expenditures versus incomes*--They are related. This applies to households, businesses, and governments.
- *financial versus real*--Financial effects mirror the real.
- *property values and damage*--Risk of earthquake damages should cause real estate values to decline. Both cannot be counted as a loss.

3. Damage Assessments Should Not Include Land *and Depreciation*. It is important to distinguish between damage to structures and a possible reduction in the value of building sites. This is clearly a nontrivial point. In parts of California such as San Jose, the value of real estate is driven mostly by location, with building improvements contributing only a small percentage of the property's overall value. Occasionally, damage assessments incorrectly utilize the total market value of a house in deriving estimates of earthquake losses. This of course is inaccurate, since the site still has value. In fact, some unreinforced masonry buildings may have negative value relative to that which an alternative structure or land use might bring. This is particularly true when rent controls and tenant laws serve to prolong the use of unsafe buildings. Failure to separate equity in the structure versus equity in the platted site may

explain why homeowner equity is an insignificant variable in the decision to purchase earthquake insurance.

4. Expected Losses Should be Derived from a Dynamic Building Stock. The inventory of structurally unsafe buildings will change with the occurrence of events. A moderate earthquake in the range of Richter 6 to 7 would render many unreinforced buildings uninhabitable. Condemnation (red-tagging) of such buildings and their subsequent demolition would reduce the stock of hazardous buildings subject to failure in the event of a catastrophic shock. By assuming that the number of hazardous buildings remains constant over time, as is typically the case, the benefit of seismic improvements may be overstated.

5. Mitigation Policy Should be Based on Avoidable, Not Total, Loss. Loss management should emphasize the trade-offs involved in purchasing additional seismic safety. All too often, policy-relevant losses and total losses are confused. This tends to skew public priorities irrationally. Events which evoke images of catastrophic damage often serve to fuel political rhetoric, but seldom illuminate those strategies which are economically efficient. For example, the widespread destruction Hurricane Hugo's winds wrought on the Francis Marion National Forest resulted in untold recreational losses, in addition to lost timber. However, from a mitigation standpoint, the value of destroyed timber and recreation days lost is of little relevance. One must ask whether the event could have been avoided at reasonable expense. The answer is no!

6. Federal Priorities Should be Based on National, Not Regional, Impacts. Loss studies are conducted for a variety of reasons, some of which are self serving and some not. Clearly, regions impacted by disaster will find it in their best interest to shade facts in the hope of triggering a disaster declaration. Although quite understandable, Dr. Cochrane is not interested in such practices. This paper focuses instead on losses viewed from a national perspective, and emphasizes in particular the use of loss data in the development of economically efficient strategies for coping with earthquake consequences.

7. Count Secondary Losses. With this in mind, one must exercise great care in the interpretation of loss data. The provision of federal aid, the receipt of insurance monies, or even bearing the loss by paying for repairs out of pocket will temporarily stimulate the affected economy, thereby masking secondary impacts. Transfers serve to simply shift employment changes to another point in time or to other regions. Therefore, although secondary employment effects have not been observed, they should be estimated and counted.

Observations about Loma Prieta

This overview reflects a set of principles which are commonly accepted, at least by academic economists. However, as in the case of Loma Prieta, they are often ignored in practice.

Damage estimates persist at $5 billion to $10 billion. This represents less than 1 percent of the region's capital stock. To put this into perspective, the loss is less than the annual rate of depreciation, or the capital loss resulting from a 0.1 percentage point change in the long-run rate of interest. It appears that secondary impacts may amount to no more than $2 billion, and may more likely prove to be half that figure. This too is a relatively small amount, less than 1 percent of the region's annual income.

In some cases, these estimates are made within 48 hours of the disaster and not subsequently revised. Except for calls from economists trying to recalibrate national economic models to accommodate the effects of the earthquake, news reporters were the most frequent inquirers about losses.

The following observations about earthquake losses were gleaned from Bay Area press clippings, discussion papers, and interviews with decision makers. They are labeled as misconceptions because they violate one or more principles of loss measurement as put forward in this paper.

1. *Lost economic growth should not be considered a cost of the earthquake.* The Federal Reserve Bank in San Francisco indicated that there might be long term-losses if prospective businesses and residents decide not to come to the Bay Area. This is not necessarily a cost. It might reflect new information about the true costs of locating in the San Francisco area.

2. *The negative effects on employment should be tracked separately from the stimulative effects of postdisaster reconstruction.* Economic disruption stemming from the earthquake was masked by outside aid (e.g., disaster unemployment assistance, Small Business Administration loans, and state and federal aid to governments). Reconstruction financed by borrowing and insurance settlements also tended to dampen secondary impacts.

3. *Damages from the earthquake are not equivalent to the market value of the property.* Even if the entire structure is destroyed, the site still has value. Be sure to account for depreciation. A new structure has a longer useful life and is safer.

4. *Accounting stance is important.* Economic impacts were uneven. Bookings at San Francisco hotels declined, while business in San Jose was unexpectedly strong.

5. *Lost sales is a poor indicator of loss.* Shopkeepers tended to report lost sales as the cost of disaster, rather than lost value added. Sales include the value of goods imported into the region, and which could be sold elsewhere.

6. *The costs of unemployment should not be added to lost revenues.* Wages are reflected in sales revenue. If it is claimed that hotel occupancy rates dropped by 10 percent, then hotel revenue is reduced by 10 percent (assuming that the rate per room remains unchanged). To then add the cost of unemployed hotel workers to the lost revenue involves double-counting; room rates include all labor costs.

7. *Lost leisure should be included as a cost of disaster.* Some analysts speculate that productivity losses were held to a minimum by a labor force

willing to endure lengthy commutes to circumvent damaged roadways and bridges. However, longer commutes imply lost leisure, which has its own costs.

8. *Retrofitting to make freeways and structures safer is not a cost of the earthquake.* The temporary repairs of $60 million to the Embarcadero, I-280, and U.S. 101 are attributable to the earthquake. (The $1.7 billion worth of seismic design changes is not.) Since the design changes reflected in higher construction costs are not a product of the earthquake, they should not be counted as a cost of the earthquake. Technically, it is appropriate to count only the cost of restoring the freeways to their prevent condition.

9. *A decline in sales and income tax revenue is not necessarily a disaster-related loss.* If damages include lost income, they also implicitly include lost tax revenue. Taxes are a function of income and spending. To include both involves double-counting.

History may be a Poor Guide-- the Great Tokyo Earthquake, 1923

Much of what has been covered reflects accepted microeconomic principles. At the macroeconomic level, however, there is much less consensus about how an economy is likely to respond in the event of a large earthquake. As a result, forecasting losses at the national or international level is still a highly subjective undertaking. There are several reasons for this. Most important, disaster scenarios typically tend to reflect the economic contraction observed in 1929. Whether such a sequence of events could be repeated in the 1990s is for some reason never addressed. Second, economists tend to focus on U.S. disasters. By so doing, they are failing to recognize that the U.S. economy is now embedded in a much larger financial system, which could be destabilized by earthquakes either here or abroad. (As will become evident in the remarks which follow, I am extremely critical of the state of the art when it comes to macroeconomic loss assessments.)

The organizers of the conference wanted the presenters to restrict their presentations to domestic events. However, the *Coming Great Tokyo Earthquake* provided a number of interesting possibilities for discussing the previously mentioned shortcomings. The following scenario will stimulate discussion as to whether the Richter 8.2 San Andreas earthquake is still the truly catastrophic event for which plans must be made.

A recurrence of a 1906-like event in the Bay Area might cost $50 to $60 billion. Would this be large enough to create serious macroeconomic effects? Maybe not. What about a Richter 8.2 in downtown Tokyo? The cost of such an event has been estimated by Japanese economists to be $60-0 billion. What would be the implications for the Japanese and the U.S. economies? There can conceivably be two scenarios, one fairly gloomy and one which produces a brighter conclusion. It is not known which is more likely to occur, but the more positive scenario is intriguing.

At the time of the event, the Japanese economy is likely to be the size of the U.S. economy, approximately $4,000 billion GNP. Currently Japan's trade surplus with the United States is approximately $50 billion annually, mostly in real estate, T-bills, and select small companies. Japanese households could alone provide enough savings at the time of the event to pay for the cost of rebuilding Tokyo. No U.S. assets would need to be liquidated.

Possible Financial Repercussions on the U.S. Economy

This conventional scenario follows a pattern similar to that observed after the 1929 stock market crash. The triggering mechanism is different, but the results are identical. According to Japanese economists, reconstruction costs will be financed by a massive sell-off of U.S. Treasury bills. U.S. bond prices plummet, driving interest rates up 5 percent. The sudden rise in U.S. interest rates depresses home buying and capital investments, which produces a deep recession, paralleling the recession experienced between 1929 and 1936. Japan, on the other hand, experiences rapid growth, fueled by internal, not external, demand. Japan emerges as a super economic power, while the U.S. economy languishes.

The Great Tokyo Earthquake will strengthen both countries. First, the Japanese economists have overstated the magnitude of the event. They have probably included land values in their estimates and other impediments which have driven up the cost of Tokyo real estate. Land will not be destroyed, only structures and infrastructure. The event will be costly, but as pointed out above, the Japanese can muster enough savings to rebuild Tokyo without selling Treasury bills. The interest-rate differential will be attractive enough to induce them to hold onto these assets. Even if they are sold, the interest-rate effects are likely to be smaller than the 5 percent increase some have forecast.

The event will provide a reason to implement the recently signed trade agreement, which emphasizes the opening up of construction markets to U.S. firms. It is unlikely that under normal circumstances this would occur voluntarily. Interest rates will likely rise in the United States, but that is a normal economic response. We might view Japanese investments in our T-bills as their form of disaster insurance. The disaster will stimulate demand for U.S. building materials, resources, and services. Building in the United States will be temporarily curtailed, but this should not create significant problems, since many regions of the United States are experiencing an oversupply of office space (e.g., Houston and Denver).

In short, the disaster might serve to rectify the trade imbalances which have dominated public concern for the past decade.

Some may question whether this latter scenario is overly optimistic. In considering recent events such as the unification of Germany, the New York and Tokyo stock market crashes, and the savings and loan crisis, it can be concluded that the U.S. economy is more resilient than many observers would have us believe. For this reason, it is probable that some postdisaster projections about the performance of the economy have been overly pessimistic.

Useful Results can be Obtained from Simple Approaches

Loss estimation is imprecise, based on an incomplete and erroneous conceptual foundation, and continues to rely on hastily collected and inaccurate data. Data for many loss categories presented in this overview are simply unavailable. The purpose of loss studies is all too often politically motivated. There are exceptions, such as FEMA's and San Francisco's well-planned attempt to deal with its stock of unreinforced masonry buildings. Methods for valuing impacts must be simple. Fewer than 10 percent of those who profess to use loss data are trained in economics, and even fewer are familiar with the principles of loss measurement as promoted in the Water Resources Council's Principles and Guidelines. Despite these impediments, a simple and internally consistent set of principles can most likely be set down to assist in the presentation of earthquake losses.

It is suspected that a correct accounting of losses would show a level dramatically different than that which has been gleaned from recent events in San Francisco and South Carolina. There are several obstacles which inhibit the profession from making inroads, the most intractable of which is its stubborn attachment to outmoded views of financial and economic systems. It almost appears as if the stock market crash of 1929, and the ensuing deep depression, has left an indelible mark on the discipline's orientation. All potentially destabilizing events tend to be cast in the 1929 mold. "The Great Tokyo Earthquake will produce panic selling on Wall Street." The rest of the scenario is identical to that of 1929 and 1936. "The Great Santa Clara Earthquake will produce a shortage of microprocessors, which will bring the nation's capacity to process and disseminate information to a halt." "The economic consequences will be as catastrophic as, if not more catastrophic than, that produced by the oil embargo of 1974."

Clearly, institutions and the economic realities of the 1990s have changed such that the mechanisms for transmitting economic shocks have been altered in ways we can only imagine. The U.S. economy is firmly embedded in the world economy; this more than anything else has changed the very nature of the catastrophic event.

GENERAL DISCUSSION OF CHAPTER 3

QUESTION: The Japanese studies generate a much larger loss possibility than Europe. They are pretty low actually. They were way up in the billions.

DR. COCHRANE: More than $600 billion?

QUESTION: It was $655 billion.

DR. COCHRANE: We are close enough, but it could be argued that these estimates might include land value. I must admit that I am unsure how the Japanese performed their analysis. However, I tend to be skeptical. My guess is that reconstruction activity would produce some positive spin-offs. Tokyo has grown in ways that have overtaxed the city's infrastructure. As far as the $600 billion estimate is concerned, it is startlingly large.

This type of event conjures up images, albeit somewhat dated, of economic and financial collapse. Those that forecast such dire consequences anticipate an event that triggers a massive sell-off on stocks and bonds, and that produces a rise in interest rates, declining investment, and lowered aggregate demand, all of which serve to induce a rise in the U.S. unemployment rate. Such projections might conclude with U.S. losses amounting to $1 trillion. This type of thinking, in my view, is unimaginative. I agree that this sequence of events did emerge after the stock market crash of 1929. But, it did not in October 1987, when the collapse produced approximately $500 billion to $800 billion in paper losses. Clearly, something had changed between 1929 and 1987. The economy is no longer responding as it once did. It is this change in response that should attract our attention, not a replay of the 1929 scenario. At this stage, we probably do not have a good understanding of the loss-transmission mechanism. But, in my opinion, the simplistic projections I have seen appear to be overly pessimistic. They overstate the economic impacts.

QUESTION: Double-counting is a real problem, but there are regional issues that must be considered. In order to understand and do something about some of these problems, it is important to measure things which--from an overall point of view--could be involved in double-counting. There is a real need to study that and also to study distribution practices. When we are doing an overall assessment of the disaster- or hazard-mitigation policies, we do one kind of analysis. But for regional and local policies or state-level policies, we have to do other kinds.

DR. COCHRANE: If you are a regional economist, you had better use an accounting stance that reflects your client's interests. On the other hand, since these presentations on national priorities, a national perspective must be adopted. Some of the issues talked about earlier illustrate this difference in perspectives. For example, negative employment effects felt in the disaster-stricken region may be offset by positive effects elsewhere. In my view, a national--as opposed to a regional--perspective is appropriate.

4

Differential Impact of Earthquake Events

The purpose of this chapter is to raise questions about the differential impacts (or effects) that earthquake events may have. Certainly, earthquakes of different magnitudes and intensities will have differential effects, as will seismic events that take place in different types of geologic areas (as were described by Dr. Hamilton in his presentation). Besides these geophysical conditions, however, it must be noted that the types of social and policy contexts in place in the communities which are struck by a major earthquake will also have an effect on the types of impacts that are sustained. These differential impacts may be related to the stage of the city's life cycle (whether it is old and aging or new and robust), the types of building stock in existence, the extent to which production is localized or dispersed nationally, the extent to which a major market is disrupted, the extent to which earthquake insurance is available and affordable, and the extent to which mitigation efforts have been undertaken.

The first presenter in this chapter is **Professor Anthony M. Yezer** from George Washington University. Dr. Yezer has a doctoral degree in economics from MIT, with an emphasis on applied microeconomic theory. His presentation will focus on measuring the effects of catastrophic earthquakes in different regions of the country.

The second presentation will be made by **Professor Howard Kunreuther** and is a joint effort with his coauthor, **Professor Neil Doherty**. Dr. Kunreuther is a professor of decision sciences and director of the Wharton Risk and Decision Processes Center at the University of Pennsylvania. In the recent past, he was director of the Decision Risk and Management Science Program at NSF. Dr. Doherty is a professor of insurance in the Wharton School at the University of Pennsylvania. The work that forms the basis for this presentation focused on the role of insurance compensation, incentive mechanisms, and regulation as policy tools to reduce the impacts of disasters. Their presentation will focus on the role of loss-reduction measures.

PRESENTATION OF ANTHONY M. YEZER

Unfortunately, there has been relatively little research on the topic of economic effects of serious earthquakes by economists. Thus, the statements that can be made are based largely on extension of economic theory that has been developed to analyze effects of phenomena analogous to earthquakes. Such applications of theory also provide perspective on the types of questions that should be asked in trying to assess economic effects of earthquakes.

First, a review of the literature on economic consequences of disasters must be conducted. Second, the issue is placed in general context of economic theory by analyzing the relationship between natural hazards and economic development of a region. A hazard event, such as a serious earthquake, has a direct and immediate effect on the capital stock of a region and on the physical health of its residents. Then there is a long-run effect that follows the event as the expectations for future productivity of the region change. It is important to consider both the immediate and the long-run effects when attempting to characterize economic effects of hazards on a region.

Third, evidence on how uncertain natural-hazard events enter an economic model of development of a regional economy is considered. Special problems arise in connection with infrequent and very uncertain events such as serious earthquakes. It is then possible to conduct exercises in which the regional economy responds to disaster events. Finally, implications for additional research, particularly directed to earthquake hazards, are presented.

Before beginning, it is important to frame the question of the Forum on Earthquake Economic Issues in economic terms. What is meant by "economic consequences of a catastrophic earthquake"? What is the alternative to having a serious earthquake in 1990? Is it having a serious earthquake in 1991? Is it never having a serious earthquake in the history of the world? Is it having two serious earthquakes in 1991 to make up for the one that was missed in 1990? The manner in which the question is posed is crucial for discussion of the problem of earthquakes and the notion of economic effects. Obviously, delay of an earthquake by one year is relatively trivial compared with reduction in the total number of serious earthquakes. However, it is not clear that we have the option of lowering the total number of earthquakes, and failure to have an earthquake this year may simply mean that it has been delayed. Economic agents are assumed to treat earthquakes as random events that occur according to a Poisson process in which the probability of having another earthquake is independent of the number of previous earthquake events.[31] The occurrence or nonoccurrence of earthquakes this year provides information that is used in updating forecasts of future earthquakes.

Previous Studies of Natural-Hazard Effects on a Local Economy

There is literature that relates the asset prices of housing in a given area to proximity to a natural hazard.[32,33] The standard finding is that houses farther from the hazard sell for higher prices, and these appear to reflect differences in insurance costs. Recently, contrary evidence was published that indicates that appraisers, lenders, and buyers appear to ignore earthquake hazards.[34] The announcement of a possible future disaster in Mammoth, California, had important effects on property values.[35] Such results suggest that there is a market response to disasters. These studies have little or no dynamic component and have been done for single areas in which endangered land could be compared with safe locations.

A 1985 report to NSF provided a brief but cogent review of earlier evidence on the longer-term economic effects of disasters.[36] The tests provide very mixed evidence on economic effects of disasters.[37] The differences in the literature appear to match an extensive list of case studies on large disaster incidents against an econometric estimate of long-run effects on housing markets and survey evidence of local officials.[38,39,40,41,42,43,44,45]

Case studies of large disaster events provide great detail and document the importance of individual area responses. They often find that the disaster only interrupts economic trends and is followed by a continuation of the predisaster economic decline or advance. In some cases, substantial changes in the growth path of the local economy occur in the wake of a major disaster. It has been argued, based on the aftermath of the Great Alaska Earthquake of 1964, that the rush of aid in response to a major disaster gives a community a chance to reverse a previous pattern of long-term decline.[43] The opportunity to rebuild on a massive scale, rationalizing the provision of public services to introduce the latest technology, could open a local economy to production possibilities that might otherwise locate elsewhere. While most case studies have shown significant long-term effects, whether positive or negative, the record also contains observations of little or no effect.[40] Overall, the case studies provide mixed evidence, at best, on local economic changes following disasters.

A major econometric study of a large national cross section of disaster events occurring during the 1960-1970 period,[44] found no long-term effects of disasters on population or housing trends. While this study has been criticized for using only population and housing units as indicators, the theoretical analysis conducted here suggests that population and housing changes could be appropriate indicators of local effects of disasters if the proper tests were performed. The same authors provide additional support for the no-effect results by conducting opinion surveys.[45] They find that natural-disaster concerns are not particularly important among public officials, many of whom might be charged with dealing with their consequences. Of course, recent occurrence of a disaster can elevate the priority of hazard/disaster concerns temporarily; however, disasters were far down the list of priorities for most officials who responded to the survey. The evidence of sensitive housing-market reaction to the announcement of earthquake risk[33] contrasts sharply with reports on the lack of long-term effects[44] and the reactions of real estate market participants.[34]

Natural Hazards and Regional Economic Development

The literature on regional economic effects of natural hazards reviewed above is not based on standard economic models of an urban region. Such models require a city to compete for capital and labor inputs against other regions.[46] It is possible to add natural-hazard considerations to the standard open-city model.[47] These models stress the effects of hazards on the supply side of a regional economy that must compete with other locations in a general equilibrium model of production.

In order to attract capital, a city must provide an expected risk-adjusted rate of return equal to that available elsewhere in the economy. Similarly, the firms in the city must be able to provide expected real wages, adjusted for differences in cost of living, which are as high as the real wages provided outside the city.

The probability of a natural-hazard event lowers expected real returns to capital and expected real wages. This inhibits regional growth. This is consistent with maximization of economic welfare in that development should not be encouraged where the probability of natural-hazard damage is high. Still, if labor and capital are extremely productive in an area, it may grow substantially even if natural-hazard events are quite likely.

Occurrence of a hazard event, such as a serious earthquake, has both immediate and long-run effects on city economic development. The immediate effect is based on destruction of capital, both industrial plant and equipment and housing. This results in lower realized rates of return on capital and lower real wages than were anticipated. Government assistance programs may raise these realized rates of return by compensating firms and households for immediate damage to capital. The relationship between payments to replace damaged capital equipment and the true measure of immediate damage (based on the difference between expected rates of return and real wages and realized rate of return and real wages) is problematical. It is possible for some individuals to be damaged substantially, through loss of expected real wages, without sustaining any capital loss that is the object of direct government compensation. Similarly, the replacement cost of business plant and equipment bears an uncertain relationship to losses in expected returns.

The long-run effects of a natural-hazard event arise because firms and workers will produce new estimates of expected returns and real wages based on experience of a disaster. Specifically, experience of a serious earthquake may cause firms to lower their expected returns to plant and equipment investment and workers to lower expected real wages, so that capital and labor are encouraged to locate outside the city.

A detailed analysis of a general-equilibrium model of a city subject to natural-hazard risks is presented in *The Local Economic Effects of Natural Disasters*.[48] While there are many implications of such models, among the most interesting and relevant for the discussion of serious earthquakes are the following. The one reliable indicator of overall effects on the regional economy of a natural disaster is the change in land values. Other seemingly attractive indicator variables (e.g., wages, population, and output) do not provide a reliable index of the shock to the economy generated by the event. It is possible to have an earthquake event enhance productivity by destroying outmoded capital stock and perhaps provide an opportunity to rationalize the use of land in the rebuilding process. This result may appear shocking to many. Only a handful of economic studies have used the land-value approach. Nevertheless, this is the approach suggested by theory. A second result is that immediate effects (i.e., an inventory of capital investment destroyed and human injury) have little relationship to the overall effect on the regional economy. The long-run effects, which may be far larger than immediate effects, are generated by changes in expectations of future productivity. Finally, a natural disaster can affect the regional economy through a number of routes, including both labor and capital markets, and a general-equilibrium economic approach is necessary to trace the outcomes.

Expectations and Economic Effects of Disasters

The primary factor generating long-run effects of a serious earthquake is the change in expectations of future returns to capital investment and real wages of labor in the region. Simple announcement of a significant increase in earthquake hazards can have expectations effects as well. Thus, an understanding of economic effects of earthquakes must consider the way in which such events cause individuals to update their forecasts of future earthquake activity.

Land and housing markets in a city incorporate an adjustment to the prevailing expectation of natural-disaster frequency. The expectations hypothesis implies that, if actual disaster rates equal expectations, there should be no significant response in the city housing market, because unanticipated disasters are equal to zero. Thus, there is a need to develop an anticipated-disaster-frequency measure in order to determine if the actual disasters were more (or less) frequent than expectations, i.e., to measure unanticipated disasters. This question of expectation formation and measurement of those expectations is essential to understanding effects of earthquakes.

The expectations hypothesis regarding market responses to disasters implies that, if the frequency of earthquakes in an area during the 1980s were identical to prior expectations, then the observed disaster rate during that period would have no effect on economic activity. Unanticipated disasters are equal to zero in this case. If actual disaster experience were significantly higher (or lower) than expectations, the expectations hypothesis suggests that disaster expectations would rise, and consequent negative effects on land rents would be observed.

For example, the occurrence of three earthquakes during the 1980s in a part of California expected to have one (three) [five] floods per decade should have a negative (neutral) [positive] effect on expectations of earthquake hazard and a corresponding positive (neutral) [negative] effect on the local economy. In an area expected to have three earthquake events per decade, the danger of earthquake has already been discounted at that frequency and is reflected in both land values and levels of employment and population. As unanticipated disasters increase from -2 to 0 to +2, the local economy experiences increasing negative effects. Note that these negative effects arise even if the community makes extraordinary efforts to mitigate disaster damage, because the expenditures for mitigation lower the expected returns to investment. Obviously, such mitigation may be a very sound economic investment if disaster expectations rise, but this merely recognizes the cost of the potential disaster event.

The expectations approach to measuring economic effects of natural disasters was tested using property value data for U.S. cities.[48] Working with flood, fire, and windstorm events, it was found that unanticipated disaster events had a negative effect on property values, while anticipated disaster events had no effect. This explains the failure of some studies to find negative effects for the total number of disaster events. The long-run effects calculated as a consequence of unanticipated disasters were far larger than estimates of immediate damage from these events.[45]

Problems in Predicting Earthquake Effects

The expectations approach to earthquake hazards suggests that the long-run effects of a particular serious earthquake event may be very substantial. However, these effects depend on prior expectations and the way in which expectations are updated as a result of the event. There has been little research into the way in which individuals modify their expectations of natural disasters in general and, in particular, of earthquakes. At present, there is no mechanism for assessing the nature of such expectations in different regions of the United States. Without such initial expectations information, it is difficult to anticipate the full long-run effects of a serious earthquake.

Research Needs

Research into the manner in which disaster expectations in general and earthquake predictions in particular are formed and the role of the government in providing an information base would be most useful. It is particularly important to determine if market responses to new information, in the form of land-value changes and mitigation efforts, are efficient. Land value studies conducted for areas experiencing disasters can allow assessment of long-run economic effects. It is not clear that disasters have permanent, long-run effects on economic growth of regions.

PRESENTATION OF HOWARD KUNREUTHER

The following presentation is a summary of the work of Ann Butler, Neil Doherty, Anna Kleffhe, and Howard Kunreuther--at the Wharton School. What are the appropriate policy tools to utilize as this very difficult problem related to earthquakes is dealt with? There really is not a very good understanding of on the risks from an earthquake. There is a lot of uncertainty and ambiguity about the probabilities of an earthquake and both the primary and the secondary losses--the economic impacts on a wider level.

The role of loss-mitigation measures is one aspect of this problem. It is a tricky problem in several different dimensions. There will always be the difficulty of going from the catastrophic earthquake to the concept of expected losses and expected benefits. Following the impacts of a catastrophic earthquake, this mitigation measure may have cost only $1,000 for a private home, yet would have saved $20,000 in losses. Then one would ask the question, What is the probability of that particular quake occurring? If it is 1 in 100, you multiply 1 in 100 by a saving of $20,000 and it becomes only $200. As a result, the mitigation measure which looked very impressive after the fact may not have been so impressive if it is evaluated before the fact.

There are three areas that need to be covered: (1) the set of loss-mitigation measures for reducing quake losses; (2) preliminary results from an interactive mitigation model that the Wharton Risk and Decision Processes Center has undertaken for the Federal Emergency Management Agency (FEMA) and Dames and Moore--It utilizes benefit-cost analysis to evaluate alternative mitigation measures; and (3) recent research on the role of insurance in encouraging households to adopt the loss-reduction measures (LRMs).

A motivating question is: Should loss prevention measures be required as a condition for households purchasing insurance? The simple answer is that it depends. What assumptions are being made regarding the choice processes of individuals? How accurate are the estimates of the risk made by the homeowners who are considering the purchase coverage, as well as the insurance industry that is setting the rates?

What kinds of decision rules are utilized by individuals? We will start off with the assumption that people are rational and maximize expected utility or expected value. In reality there are many other decision rules that people use in making these decisions. That will have an enormous impact as to whether mitigation should be voluntary or a condition for insurance.

Another aspect will be, of course, the nature of the insurance program. Is it voluntary? Is it required? How are the rates going to be set? What is the level of coinsurance, deductibles, and all terms of coverage? There are a whole set of externalities associated with disasters that need to be considered in determining the role of mitigation measures. Lives lost, injuries, business interruption, irreplaceable objects, and the impact on the community of a catastrophic earthquake come into play, of course. What is the nature of disaster relief, if one is not protected if these mitigation measures are not taken? Will government come to the rescue and bail the victim out? Of course, if the victim really believe that someone is going to bail him out, then there is no reason that he should take action before the fact.

Now, a question that could then be raised is, what is the role of loss reduction measures (LRMs) in dealing with an earthquake program? There are a number of obvious ones. It will reduce the physical damage from the earthquake, and the number of injuries and the lives lost, as well as the direct and secondary losses discussed in previous chapters. The workers' compensation, mortgage default, temporary housing, or business interruption may be spared if mitigation measures are adopted and some of the damage is prevented. It is hard to determine what mitigation will do to one's personal insurance rates, for example. Of course, it reduces the need for disaster relief.

The following is an interactive model that gives an idea as to how this problem is viewed from the point of view of expected benefits and costs. The assumptions may be tricky, but the model is very simple. Consider a particular earthquake. What is the probability of that earthquake occurring? The damage that would have occurred with and without the adoption of mitigation must be multiplied by the estimate of probability to determine the expected benefits of mitigation. If there is a very large saving from mitigation but a very low probability of a catastrophic earthquake, then the expected benefits from mitigation may be relatively small. When the benefit-cost analysis is utilized, a very small number for expected benefits may result, even though an earthquake occurs, and the reduction in damage could be enormous.

There are a number of interesting issues that affect the way one evaluates the performance of mitigation measures. One of them has to do with the discount rate. If the losses are discounted by a very large discount rate, then the expected benefits are going to be much smaller than if a smaller discount rate were utilized. Mitigation measures are paid for when they are adopted, but the benefits accrue over the length of life of the house. If the future is discounted by a very large interest rate, a relatively small benefit results from the mitigation measure if an earthquake occurs. Think, for a moment, 10 to 15 years ahead of time. Should the earthquake loss occur 50 years from now, it is going to be viewed as insignificant in present-value terms.

Another issue is the sensitivity by different stakeholders to different losses, if it is the direct losses you are talking about. What impact will the mitigation measures have on the property owner, the insurance industry, the general taxpayer, the developer, and the real estate agent? A lot of these interactions that have occurred between stakeholders is why we have a very fascinating political problem today. Certain parties are going to benefit and others are going to lose from specific programs.

As a part of the interactive model, we undertook sensitivity analyses with respect to the discount rate and the types of losses. The situation where the benefit-cost ratio will not be as attractive for any mitigation measure is when you have a very large discount rate--for example, 8 percent--and the damage is restricted to property. If the benefit-cost ratio is greater than 1, in this case, the mitigation measure is an attractive one.

On the other hand, suppose you utilize a zero discount rate over a 50-year period and you take into account secondary losses. If the benefit-cost ratio is less than 1, then this mitigation measure should not be adopted.

Now consider the case of a decision maker, like a homeowner, who has to make a decision whether or not he wants to adopt a mitigation measure.

The reason, of course, that this is critical is that in some sense a lot of the questions that are asked are, where do we let people voluntarily take steps? where are building codes? and where is insurance linking up with mitigation measures? These are all open questions.

The first assumption is that the homeowner is risk averse. The whole theory of insurance assumes that people are buying insurance because they are averse to risk. They want to avoid a catastrophic loss by paying a small premium. People are assumed to be rational, and they maximize their expected utility.

A second assumption is that the loss-mitigation measure is amortized over a period of time, 20 years. The reason for that is because there is a very large up-front cost, but there is a benefit that accrues over a long period of time. So the mitigation measure will be broken down to an annual cost, even though in reality all the money would be spent up front. The mitigation measure will impact on the magnitude of the loss and/or the probability of the loss of a given magnitude.

A mitigation measure could be thought of as reducing the damage, from $50,000 to $20,000 for a particular quake that would occur. Or it could be viewed in a different way. For example, what is the probability of a quake causing a certain amount of damage, like $50,000? If the mitigation measure, was imposed that probability might go down to zero, or it might go down to a much lower number than $50,000. A mitigation can be perceived as wither a loss-reduction measure or a probability-reduction measure or both.

It does not seem like an important assumption to make from the point of view of what will be covered on benefit-cost analysis. It makes a big difference analytically as one begins to look at the model. Benefit-cost analysis will be discussed later in this presentation. It is something that has come out to be a rather important element in terms of judging what is good and what is not so good from the point of view of benefits.

The following four considerations will guide the analysis:

1. *Myopic behavior.* It is recognized with mitigation measures that cost is borne once and benefits accrue over time. Immediate costs can be focused on, so if asked whether one wants to adopt a mitigation measure or not, one would say this measure will cost me $500. Someone says, you are going to live in that house for 20 years. Why not discount it? why not look at all the benefits over this period of time? The response would be the same, the cost is $500. That is what must be paid. There is no interest in what is going to happen 10 years from now. If people behave this way, then a lot of the assumptions that have been made on the expected benefits may have to be modified.

2. *Misperception of the probability.* This earthquake is not going to happen to me. Why should it happen to me? I am living in an area, and we had our quake a year ago in San Francisco. I am certainly not going to think about an earthquake happening in the next 10 years. If it is thought that the probability is extraordinarily low, mitigation is not going to be an appealing tool. Why

worry about making an improvement on my house when the earthquake is never going to happen?

3. *Availability of disaster assistance.* If disaster relief funds are available from the federal government, why improve the house? In the past, the federal government has come to the rescue, and people may expect them to do so again if there is some kind of catastrophic event.

4. *Impact of insurance coverage on adoption of mitigation measures.* The first factor is whether or not one has risk-based rates or rates that do not change as a function of mitigation. The second factor has to do with how insurance impacts on risk aversion. If insurance is purchased, interestingly enough, there is less incentive to mitigate, everything else being equal. If insurance is purchased, the purchaser is protected with a small premium against a very large loss. What is being done through mitigation? The purchaser is protected with a small payment against a larger loss. By taking out insurance, the aversion to risk has already been reduced. Therefore, mitigation in and of itself is a less attractive option than if the insurance had not been taken out.

For example, there is an earthquake of intensity 8.0, with the probability of .0181. The home value is $100,000, and the damage without mitigation is $11,000. The damage with mitigation is $4,700. The cost of mitigation is amortized on an annual basis to obtain $111 a year, and the benefit of mitigation is $114 per year amortized, so the benefit-cost ratio is 1.03 on these figures. The premium on an insurance policy will be determined by the probability of the earthquake. Notice that the focus here is on just one catastrophic earthquake.

Two different cases present some qualitative results:

- In the first case everything is accurately estimated. The probability is accurately estimated by the individual, and the question is, what is the benefit of taking out mitigation measures or not taking out mitigation measures? One benefit is no insurance, the other is full insurance. Mitigation comes out best whether one has no insurance or full insurance as in this case. The time that mitigation is most likely taken is when there is no insurance.
- The second case is one where there is misperception of probability, and it is still thought that there is a chance of an earthquake, with the higher probability of .06. In this particular case, when there is no insurance, no mitigation is better than mitigation. One would say, why mitigate the problem when its probability is so small? I would rather take the loss if I am going to measure it that way. If you are forced to buy insurance, then it turns out that mitigation is better than no mitigation if you have risk-based rates. The risk-based rates are based on the true probability rather than the perceived probability, and the benefits received from mitigating are going to be greater than the cost.

Several policy questions can be explored. What factors lead individuals to voluntarily adopt the LRMs? That is an interesting behavioral question where there is some insight in terms of what people are likely to do. Some of this analysis may shed some light on what people could be encouraged to do, but a lot of this has got to do with how people perceive the probabilities, how they perceive the losses, what incentives can encourage people to take action, and when will regulation be needed.

There are some interesting questions in terms of differences between floods and earthquakes. Currently in Congress there are some questions as to whether earthquakes should be treated in the same way as floods have been. There may be some similarities, but there are also large differences between floods and earthquakes. One has to figure out how that is going to impact on the proposed program. The carrot-and-stick approach is appealing, but the question is, what kind of a carrot and what kind of a stick should be used in this situation?

GENERAL DISCUSSION OF CHAPTER 4

QUESTION: Dr. Kunreuther, how does the individual's life expectancy comes into the picture? Why worry about what happens 100 years from now

DR. KUNREUTHER: That is an interesting question, but is it a probability of .01 or is it that the earthquake is 100 years from now? As we know, it makes a real difference in terms of how one interprets it; and, certainly, if one thinks that it is 100 years from now, then you are right. At that point the probability is zero for the next 99 years and 1 for the hundredth year, if you really interpret "100 years from now" in that kind of a context. Of course, nothing matters then, just live happily ever after.

QUESTION: Your least favorite point is that individuals behave rationally. Do you find they behave more rationally when it comes to material things--such as a house, a car, physical damage--and less rationally when it becomes their perception of personal risk? I think of this in relation to the nuclear power debate, where the material risks are relatively negligible but the personal, perceived risks are enormous.

DR. KUNREUTHER: Right. Let me give you an overly simplistic answer with lots of grayness. There seem to be two kinds of events that individuals deal with, and they are dealt with differently. They are the events where we say, "It cannot happen to me." This occurs partly because they are voluntary; it is our decision. These are the kinds of events that we really prefer not to worry about because if this Pandora's box of horrors is opened, we start worrying about more and more things. With respect to natural disasters in general, no one is to blame. There is a tendency to think about these events by saying that the probability is very low or I am not going to worry about the consequences.

The contrast with the nuclear power, hazardous waste, and NIMBY problems is extremely important, because here we almost take the opposite tack. We say, "It *can* happen to me;" and, even if the experts are going to tell us that the probabilities of these events are extraordinarily low, we do not necessarily trust the experts. There is a lot of ambiguity associated with these issues and, most important, we can blame someone else for the consequences. Therefore, we do not want it in our backyard, even if people are going to say that there are real benefits.

I think one of the challenges in this whole area is to start categorizing human behavior as we look across these events. One particularly intriguing event that has not been fully studied is radon. Radon is a case where you cannot blame anyone for the damage, and yet it may be in your own house. It is not at all clear what kind of decisions people will make in terms of mitigation, for example, because they should be willing to take some action to get rid of the radon; yet everyone is saying, "I do not want to know whether my house has radon. I do not want to have a test. I do not want to have anything to do with it simply because if I know, I have to do something, and I am not sure that I want to know."

QUESTION: When considering the process of the lack of attention to mitigation in public buildings, shouldn't the function of the building be taken into account? For example, a warehouse and a school have very different functions.

DR. KUNREUTHER: That was one particular example and a particular kind of mitigation measure. We should not in any way generalize from this one case of one public building, but the point you are making is very important. It was just an illustration that there could be some mitigation measures in some situations which you might not want to undertake. What should come out in any analysis is, that a public building has lots of secondary impacts, and there are lots of benefits when all of those things are examined. In most cases, mitigation measures probably will be very beneficial.

QUESTION: Dr. Yezer, how does this relate to your concept of a public building and its role in secondary impacts?

DR. YEZER: Ultimately, we pick up the public buildings and the land value. Presumably if a park is destroyed, the land value around the park obviously is reduced. That is how it is incorporated in the model which I presented (Chapter 4).

In addition, the issue of how people react to probabilities of disasters is interesting. People assume that individuals do not react to probabilities well. If that is true, they should overestimate probabilities of future disasters right after they have just had one and underestimate them when they have not had one for a period of time. That would suggest that there is gain to be made by buying and selling houses, that firms should be ready to rush into areas that have had disasters and buy up all the houses at very low prices. I have heard of some of that behavior; but while people may say all sorts of irrational things about their expectations, they refuse to conduct a fire sale of their home. At least the market has not found them willing to do so, therefore you have to watch out for the difference between what people say and the actual behavior of the individuals in moving or in selling an asset.

QUESTION: Dr. Yezer, my question relates to economics. Even when the probability of a disaster increases, there is no apparent decrease in property values.

DR. YEZER: I was trying to illustrate in that particular case, you can have the same effect. You have the same exact physical event. In one case, you observe a lot of buildings falling down, and in another case you do not observe any falling down, because you spent all the money on mitigation. Of course, the key is that the expensive mitigation has been capitalized in the land value. In fact, you would have an unambiguous indicator of the effects of earthquakes if you focused on the right variable--land values.

It is interesting that everybody wants people to mitigate, and they are focusing on all of the observed dollar flows that they can find if there is a disaster and there is no mitigation. That is a perfectly sensible thing to do, but one must realize--just as in the case of highway accidents--that we do not want to reduce fatalities to zero, that we really do not want to have the country spend so much of its GNP on mitigation that we reduce hazard damages to zero.

QUESTION: My question is related more to perception of risk and how large a factor disaster risk is in market forces or real estate values.

DR. YEZER: Well, I would bet that property values do not fall by more than 1 or 2 percent as a result of a Loma Prieta-size earthquake. But even 1 or 2 percent of $50 billion in land values in the Bay Area is a nice piece of change. As a matter of fact, using the sort of approach I have advocated, you get much bigger effects of unanticipated disasters than you would ever get by counting the buildings that actually fall down.

QUESTION: What is the role of insurance in mitigating secondary impacts. Specifically, what were the economic effects of Loma Prieta and what role did insurance play versus the economic impacts of, say, Hurricane Hugo and the role of insurance in that economy.

DR. KUNREUTHER: That is a very interesting question because the question is really, what kind of a policy is one talking about from an insurance point of view? If you are referring to an earthquake policy that is related to property damage, that is one kind of an insurance policy. If you are referring to a business-interruption-risk policy, that is another kind of a policy. Certainly we could consult with people who are much more knowledgeable than I am about some of the details with respect to various insurance. The question that we would raise would be, are you attributing essentially the losses that come out of the quake to the right kind of policy or are you subsidizing across policies? For example, if it is life insurance, a quake policy is not really protecting yourself. By mitigating, for example, you may be able to save lives. So in some sense, through mitigation, you are cross-subsidizing life insurance and likewise with business-interruption risk, if that happens to be a separate policy.

There are two issues. The first issue is: What is the function of insurance? This is a question of whether something is insurable. Personally, I would want to turn to the people who really are marketing these policies to make some comments on that issue. The second issue is: Are the rates really reflecting the risks or is there some kind of cross-subsidization across risks?

QUESTION: This was a question for Dr. Yezer. As I understand it, the indicator in your model is changes in the values of land. If that is so, then how would you explain that around 25 years ago, when identification of the earthquake threat significantly increased, there was a trend of increasing land values in Los Angeles, San Francisco, and Tokyo, which are the highest land values in the world?

DR. YEZER: Well, other things have been happening in those areas, and you have to try statistically to hold all the rest of that constant. I can get a measure of the rate of appreciation in values of houses that was anticipated before a hazard effect, then I observe what happened afterwards and try and explain the difference. Expected appreciation rates before any hazard events in California, they are very high. High rates of appreciation occur in high-risk areas, but the only question is the partial impact of unanticipated-hazard events on these property values.

There are a lot of anticipated-hazard events in California. For example, everybody knows that if you have a heavy rain, a lot of houses are going to slide. And, when you have a heavy rain periodically and those houses slide, that does not really have any effect, because that is an anticipated disaster.

QUESTION: I am just wondering about this issue of trying to assess losses and damages. The smaller the region defined, the more the impact. For example, in the Marina District there was an incredible impact. In San Francisco, houses which are on bedrock are appreciating, and the value of land and property which are not on bedrock has gone down. There was a very significant microdifference in land value, in land prices. That is a very important issue.

DR. YEZER: You have to take this across a labor market area. Charles Scawthorn wrote a paper on what is called the "tilt effect" in the Journal of Urban Economics. If one area of Tokyo has an earthquake, then overall land values will go down; but in some parts of the city that did not have damage, land values will go up. He has modeled this effect by taking a full labor-market area in order to account for this tilt. If a flood occurs, then the property values in low-lying areas will tend to go down, and the high-lying areas will tend to go up. But overall, land prices will go down if there is an unanticipated disaster.

QUESTION: However, in your model, it seems you are assuming that every structure is going to have a loss at some point. I think that over time--say a 100-year period--there will still be houses or buildings. Was that taken into account in your process?

DR. KUNREUTHER: I presented a representative structure. As you begin to start doing this on a more systematic basis by going across structures, you would obviously have to tailor the analysis to the individual buildings. The idea basically was that there was a probability of a particular quake occurring, and then one would ask: If that quake occurred, what is the likely damage to a particular structure? As you move that structure away from the fault line, if it were a wood-frame house rather than a concrete block, you would have different kinds of damages that would take place for any kind of an earthquake. It is really a generic treatment, much in the spirit of probabilities toward losses. To the extent that you were able to look at the characteristics of that structure--and, of course, that is the reason why loss-mitigation measures get considered in the first place--you might be able to say that the damage that would be sustained by that structure would be very, very small, even for a very large earthquake.

QUESTION: If you assume a value of $100,000 for a house, the societal cost would require that we account for the whole community, and not just for individual people.

DR. KUNREUTHER: I think that is one of the interesting questions. When thinking about mitigation or kinds of policy tools, we must consider our society in various situations. We have a very hard time trying to measure the specific characteristics of these structures. This is similar to the problem we face in the insurance industry in general. Since you cannot get information on every little characteristic of each person, rates are set for a particular class. The individual may feel discriminated against because they say, "I know I am better than the average driver." Yet the principle exists that you have to establish a policy upon which a rate is based.

With respect to earthquake insurance, it would be preferable to tailor rates on the basis of knowing about each specific structure. But if it is going to cost too much to do it, we may not be able to base rates for every individual structure.

DR. YEZER: There really is an interaction between what Dr. Kunreuther is doing and what I am doing. If you get risk-based insurance premiums, then people have a basis for judgment about allocating their resources, both capital and labor. You can provide a tremendous amount of information about risk from the insurance premium. Providing information to people allows them to respond efficiently. Risk-based insurance premiums could give you a lot of information without having to go out and read books about geology and earthquake probabilities.

DR. KUNREUTHER: We also have in common a Chicago model as a part of our orientations, recognizing that it is limited in terms of what it can do. Also recognizing that things like incentive systems are voluntary and more desirable than requirements. But there may be situations where you have to have requirements. Risk-based premiums would be an example of a real incentive to give people information, an instance where you do not have to make the requirement. Will they behave in a way that takes into account that information? That may be another question.

QUESTION: There is a cost of risk-hazard mitigation. Who has to pay that cost?

DR. YEZER: Those who benefit from mitigation should pay. Transaction costs of enforcing mitigation measures is an important issue. Who pays for this? Is there a bill of sale or some kind of a contractor's bill that can be taken at face value, to prove that someone has done something to make their building safer? How can the effectiveness of mitigation expenditures be assessed? What are the mechanisms that will be used, and how will they be implemented?

QUESTION: Is your selection of 20 years for this kind of mitigation completely arbitrary? Why would it not be the life of the house?

DR. KUNREUTHER: Purely arbitrary, because we like the number 20. I think it is important that you bring it up, because we had to pick a figure, so we picked 20. We could have picked 50, we could have picked 80.

5

Resource Shifts Following a Catastrophic Earthquake

This chapter focuses on inter- and intra-regional shifts in resources following a catastrophic earthquake. What is known about the conditions under which, and the likelihood that, those kinds of shifts in resources will take place? What kinds of problems are caused by those shifts or a lack of them? What types of theoretical and empirical models exist to anticipate whether such shifts will take place?

The presentations in this chapter will raise the issues of substitutability of products and the redundancy of services as they relate to different economic sectors. Although major consideration should be given to the commercial sector, some consideration should also be given to the energy, fuels, and utilities sector. A major disruption in "industrial-strength" power, for example, may have major consequences for the recovery of heavy industry across a distribution system.

Ronald Eguchi is a civil engineer and an associate with Dames and Moore in Los Angeles. Mr. Eguchi has broad research experience in risk analysis, earthquake engineering, and natural-hazards engineering in general. He is a member of various professional societies' technical committees on lifeline systems and performance in earthquakes. His presentation focuses on lifeline systems and the major issues associated with regional shifts of resources following a catastrophic earthquake.

Dr. Tapan Munroe has a doctoral degree in economics from the University of Colorado. Since 1984, Dr. Munroe has been the chief economist for Pacific Gas and Electric Company, headquartered in San Francisco. His presentation will address the economic impacts of the 1989 Loma Prieta earthquake on the Bay Area.

PRESENTATION OF RONALD EGUCHI

This presentation focuses on the lifeline problem and some of the major issues that are important from the standpoint of the recovery of these systems after earthquakes. Direct and indirect losses have been topics of discussion in previous chapters, but nowhere does the problem of indirect losses become more significant than when lifelines are examined. For example, the direct economic effects associated with lifelines (i.e., direct damage) are small when compared with the effects associated with the disruption of lifeline services. Therefore, focusing on indirect impacts of lifeline failures is very important.

This presentation will focus on the resource problem and more specifically, how intra- and interregional shifts in resources can improve the recovery of a community after a catastrophic earthquake. Very little attention has been given to this area. Some of the major problems associated with resource shifts will be identified and some positive aspects of resource sharing will be discussed. An example of how one might incorporate seismic vulnerability studies into looking at this problem of postearthquake recovery will then be provided.

It is important to view the current state-of-the-practice for lifeline seismic design in relation to current standards. From the standpoint of seismic vulnerability studies, scientists have come very far and have actually done quite well in assessing the seismic vulnerability of various kinds of lifeline systems. A lot of work has been done on water, natural gas, and electric power systems and, to some extent, communication systems. These kinds of methodologies have been applied in different regions of the country. For example, in California systems, vulnerability studies have been applied very heavily. These kinds of studies are also being performed in Seattle. In addition, earthquake awareness is increasing in the Midwest. These studies are funded by both government research projects and some commercial clients. Several businesses are performing these kinds of studies to evaluate what their vulnerabilities might be during a catastrophic earthquake.

Planning efforts in the postearthquake response and recovery area have been limited. The primary problem is that the results from these seismic-vulnerability studies are not being integrated directly with response-planning efforts. Once this kind of integration is emphasized, certain shortfalls in inventory and response capability will become evident. Based on these shortfalls, other resources can be examined to handle the problem.

In terms of resources, this is important at three different levels:

1) supply--for example, suppose natural gas or oil supplies are disrupted, or local production facilities impacted; are there other sources that can be used to provide this supply?

2) response--in the event of a disaster, do resources exist that are needed in order to detect damage and finally isolate this damage so that it does not become a problem?

3) recovery procedures--are there repair inventories or the manpower and equipment to make these repairs? Do resources exist to bring damaged lifeline systems back up so that they become functioning systems within the community?

What are some of the major problems associated with intraregional transfer of resources? First, if a very large earthquake occurs it is very likely that similar utilities (i.e., utilities that may be relied on for resources) would also be affected, and this has happened in past earthquakes. In order for this type of lifeline (i.e., underground) to function or to operate, whatever damage has occurred to the distribution system must be repaired. In most modern earthquakes that have been examined, there is some type of distribution pipeline damage, and this damage must be repaired before the system can be operated.

There is the interutility or interaction problem that we are beginning to recognize as being more important. That is, lifelines by themselves are not independent. Lifeline systems will depend upon other lifeline systems. As in Loma Prieta, many of the water systems experienced minimal damage to the piping systems. However, because there was no electric power to run the water pumps, the operation of these water facilities was essentially curtailed. Thus, the interaction problem is important.

The interaction problem also becomes important when failures in common utility corridors are examined. It is quite possible that damage from one lifeline system may actually impact the operation of others that are adjacent to it.

The demands on local contractors and those who can supply equipment and manpower are great after a major earthquake. That is, there may be a limited number of suppliers who provide this service. The following example suggests that supplies available during normal accidents are not necessarily adequate for earthquake conditions. The neighboring utilities that may also be damaged will require immediate inventories, resources, and manpower. The fact that they require the same services will impact the repair and recovery period for the region.

So these are some of the major problems seen with intraregional transfer: specifically water, gas, and oil systems. Transportation systems have other very special problems, and interregional transfer issues become very important. If there is a major disruption to a transportation system, importing supplies from out of region or out of state will be very difficult. In addition, a lot of the equipment that is needed to make these repairs are not small items or simple things to transport, so there should be concern over transportation disruption.

Major disruption to local lifeline systems may also have an impact on other regions. For example, this may be particularly true for the Midwest when gas and oil systems are considered. A major disruption caused by an earthquake in the Midwest will likely cause shortages of supplies to the Northeast. This question of economic impact, in fact, is being addressed by the National Center for Earthquake Engineering Research in one of the studies that focuses on oil pipelines i.e., what are some of the socioeconomic impacts of a major disruption of oil supply?

Importation of interregional supplies must use existing pipeline systems. It is not as though you can change the routes or incorporate new pipeline components into the system. In order to transport these bulk supplies, they must be transported through the existing lines. If the existing lines are damaged, the supply from these different regions will be hindered.

Finally, this may not be a problem, but in the Loma Prieta earthquake last year, manpower resources were transferred quite effectively. However, if manpower resources are received they must be effectively managed and there must be places to put these people and things for them to do. If these tasks are not performed, then this may create significant logistical problems that we may not want to deal with.

There are, of course, some very positive aspects associated with the transfer of resources, particularly if a small or moderate earthquake occurs. In the intraregional case, neighboring utilities will provide a very effective resource for equipment, manpower, and possibly supplies. There were examples of this during the San Fernando earthquake, where temporary waterlines were laid out for the city of San Fernando.

Very important agreements, like mutual aid agreements that would be useful in response and recovery efforts, will likely be in effect for utilities that are located very close together. This type of planning exists, at least in California, and these agreements should facilitate any response or recovery efforts during the event. Some of these utilities will have already been involved in exercises or planning efforts related to response and recovery. Therefore, at least a vehicle or mechanism is in place to examine this problem.

Then finally, similar utilities will generally be regulated by the same agency, and to the extent that regulations and standards play a role in terms of design, construction, or repair of these types of systems, this may help facilitate a recovery effort.

There are positive aspects associated with the interregional transfer of resources. The obvious benefit is that there is a much larger supply of inventory, manpower, and equipment to pull from. The second item may not be as important or as significant as the first, but when out-of-town or out-of-state utilities become involved with a disaster, they are not so much involved with the response efforts but are more involved with the recovery aspect, and to the extent that the recovery effort is more easily handled, this may be an advantage.

Focusing on supply during a moderate earthquake, both inter- and intraregional resources may provide a reasonable source. When the magnitude of the earthquake is increased, however, to the extent that larger regions are affected, these resources become less effective.

Interregional resources can be reasonably effective, however, they will be very effective during moderate earthquakes because of their close locations and commonality of the systems. The effectiveness of intraregional resources stays about the same with large earthquakes, but the interregional resources become less effective. From the standpoint of recovery, the same kind of pattern may result. Again, the important point here is that when the earthquake becomes too large, it begins to impact the immediate resources that would be very effective during a small or moderate-size event.

Vulnerability studies have been integrated into the response/recovery planning that is being performed for a water company in Southern California. The types of facilities being considered are water wells. If a very large San Andreas event exists—an event that would break three of the major segments on the fault—it is likely that the major aqueducts coming into the area will be severed. It is estimated that the outage may be as long as 4 to 6 months. The seismic vulnerability of these facilities reveals that they will also be impacted.

The kinds of models that were developed for lifeline studies are very similar to the kinds of models discussed by Don Friedman. A hazard model is needed, that is, some way of representing what the shaking effects might be throughout the region.

It is very important to consider all the effects that might impact the performance of underground pipeline systems. Ground-failure effects (i.e., fault rupture, liquefaction, and landslide) are very important in assessing the performance of underground systems.

There are various ways that water wells, including local water-production facilities, can be made inoperable and in terms of modeling vulnerability, these various failure modes were examined. A number of these failure modes can be combined because of the similar impact that they would have in terms of repair or the failure of the facility. For example, sewage contamination and chemical contamination basically lead to the same impact or result. Sanding, casing damage, and pump motor damage all imply that the casing will be damaged and that something needs to be done about the casing to repair that facility. Wellhead damage and connecting piping damage would be things that would affect aboveground components. Finally, failure of the building enclosure may also be a reason why this facility may not operate after an earthquake.

Other factors, such as power outage, also are very critical. In our analysis, each of these different failure modes is taken and the important parameters are identified. Some of the important parameters included the type of drilling method that was employed to construct a well, for example. In certain cases, in terms of contamination--which is what we are looking at here--the voids on the sides of the casing may allow contaminants to penetrate the soil. Of course, the depth of the well is also important. The important point to focus on is the impact these parameters will have on estimating the probability of contamination at a site. The loss, if it did occur, would be 50 percent of the replacement cost of the well, and the outage time for repair could be as long as 6 months or greater.

Out of 3,000 wells about 500 of them would be expected to experience moderate to major damage. Out of the 500 approximately 166 would suffer major down-hole damage, 41 would experience minor damage to the casing, and over 400 would suffer damage to the pump motor.

In terms of impact to the region, production shortfalls would be created. When the time required to repair each of these facilities, and the expected loss are considered a time line can be developed to show what the immediate production shortfall would be and how the system or facilities would be restored as a function of time.

Based on a survey conducted, the repair resources are quite limited. Drill rigs—which are important for making these down-hole repairs—only number about 30 to 40 within the basin. Therefore, considering the total number of damaged wells--about 166--only a quarter of them can be repaired at a time. This means that a 6-week time frame may be extended by a factor of 4 when addressing the resource problem.

The same problem with resources also occurs if one looks at minor casing damage, replacement of pumps, etc. This type of information and this kind of analysis can be used to determine whether there are adequate resource capabilities. This analysis also gives a good idea of what to expect in terms of damage or vulnerabilities to lifelines. This type of analysis has been done for pipelines and should be performed for other lifeline facilities that are critical for regional response and recovery.

PRESENTATION OF TAPAN MUNROE

A lot has been presented about various types of economic impact projections. This section aims to reveal some of the specific effects observed in the San Francisco Bay area following the Loma Prieta earthquake. Some of these facts are important from an economist's point of view.

This report contains a lot of discussion about the differences between national, state, and regional perspectives. California accounts for about 14 percent of the national economy. If a major earthquake occurs in California, this could certainly have a major impact on the U.S. national economy.

The city of San Francisco was well prepared to cope with the immediate effects of the earthquake, as was Pacific Gas and Electric Company (PG&E). As a matter of fact, several months before this earthquake, PG&E went through a simulated exercise involving recovery and restoration of services in San Francisco. That experience was absolutely providential, because the speed at which PG&E could respond to the disaster, particularly in the Santa Cruz area and in the Marina District of San Francisco, was remarkable. Gas pipelines were repaired--that would have usually taken 6 weeks to complete under crisis conditions--in 2 weeks.

Pacific Gas and Electric Company was interested in assessing the impacts of this earthquake from several different perspectives: regional, intraregional, and specific industries or economic sectors. Damage estimates prepared by analysts immediately following the earthquake were problematic, with losses ranging anywhere from \$5 billion to \$7 billion. The initial estimates were much larger, about \$10 billion. "New" damages are still being discovered from the earthquake in what has become a long, drawn-out process. Our initial impression was that the loss and damage estimates were on the high side.

If one looks at the spatial distribution of the damage, it was fairly concentrated on the Marina District in San Francisco, downtown Oakland, the Bay Bridge, the Cypress section of the I-880 freeway, the Embarcadero Freeway, and Santa Cruz. The Cypress structure does not exist any longer. The off-ramps of the San Francisco Bay Bridge, the Main Street exit, and several other exits are still not repaired. It is very likely that the Embarcadero Freeway is not going to be there much longer. Most likely, there will be a sunken freeway in its place. Some of the infrastructural effects are long lasting and very likely permanent.

Significant structural recovery has occurred in a short time in the Marina District, but the effect lingers on in its economy. In Santa Cruz, there are major impacts in terms of infrastructure, buildings, and the economy. On the whole, the damage and the distribution of damage was fairly focused and concentrated in San Francisco, East Bay, and the Santa Cruz area.

In assessing economic impact in the 7-county Bay Area--using employment or income data--we find that the impact has been minimal and barely observable in the indicators (Table 5-1). Within 6 to 9 months after the Loma Prieta earthquake, economic effects appear to be nearly nonexistent. The data in Figure 5-1 is a composite index number involving six different variables, including real estate, the services sector, the retail sales sector, tourism, trade, and manufacturing, and it remained remarkably flat in the entire period. The tourism sector shows a noticeable effect. After a decline, followed by recovery,

TABLE 5-1 Economic Impact of the Loma Prieta Earthquake

	OCT		NOV		DEC	
	88	89	88	89	88	89
San Francisco Area*	3.2	2.9	3.2	2.6	2.8	2.8
Oakland Area**	4.2	3.7	4.3	3.5	3.7	3.4
Santa Cruz County	5.1	4.7	6.4	6.3	6.6	6.5
San Benito County	10.1	9.3	12.8	12.5	12.7	12.4

* Includes the counties of Marin, San Francisco, and San Mateo.
**Includes the counties of Alameda and Contra Costa.

SOURCE: Effects of the October 17, 1989, Earthquake on Employment, E.D.D., California, February 1990.

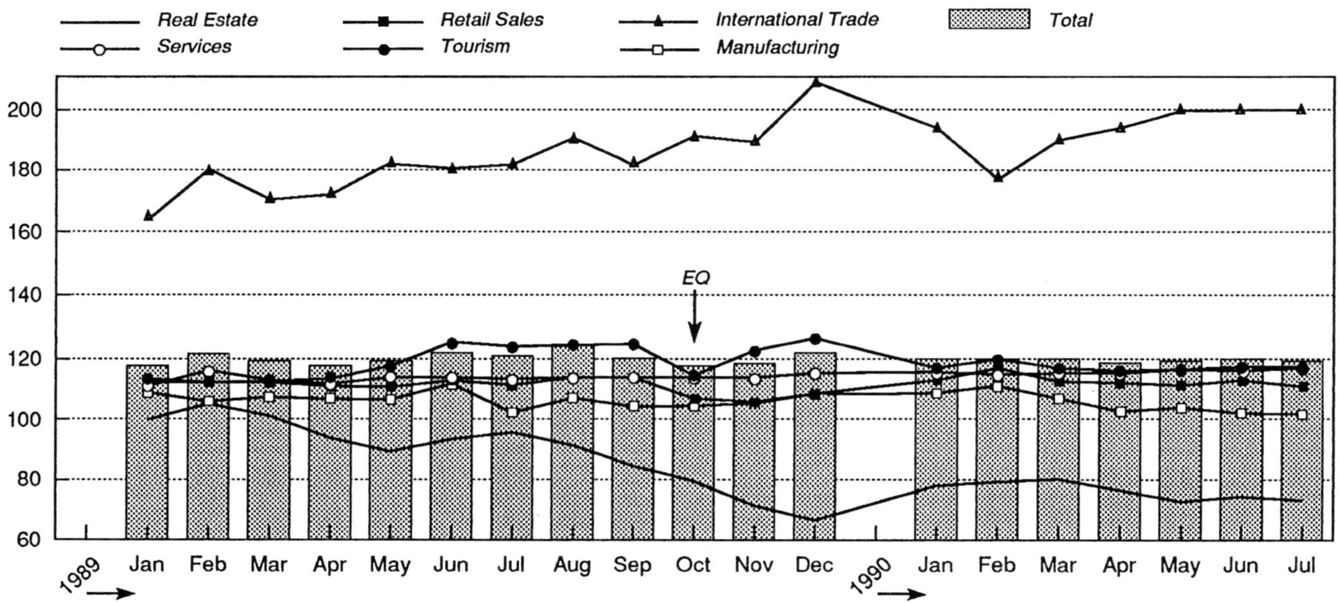

FIGURE 5-1 San Francisco Bay Area economic indicators (1985 = 100).

the indicator has leveled off. The service sector is the most stable part of the index number. The real estate component was on a decline between June and December. Of course, one has to keep in mind what was happening in the Bay Area in terms of housing affordability in this period. Only 10 percent of households in San Francisco could afford a median-priced home in the region, which partially accounts for the decline in home prices and slowdown in the housing sector. The earthquake was very likely the other factor underlying the slowdown in the housing sector.

The only sectors where some impact from the earthquake is seen are in retail sales and tourism; surprisingly, there was very little impact on manufacturing. Few production facilities were disrupted for more than 3 to 5 days. There were two or three production facilities companies in East Bay where production stopped completely, but these companies found other facilities within a matter of weeks. The manufacturing sector of the economy recovered shortly after the earthquake.

With respect to joblessness, four Bay Area counties (Alameda, San Francisco, San Benito, Santa Cruz) and two counties outside the Bay Area had most of the unemployment claims resulting from the earthquake. Comparison of the same 5- to 6-week period from 1988 to 1989 reveals that this was a short-lived phenomenon. Examination of monthly unemployment figures between October and December 1989 in three Bay area counties (Marin, San Francisco, and San Mateo) indicates that the rates are surprisingly low and about the same for all counties: October = 2.9 percent, November = 2.6 percent, and December = 2.8 percent.

Unemployment levels in the San Francisco Bay area are usually among the lowest in the country. We did not see a rise in unemployment resulting from the earthquake. If anything, there was a dip in the unemployment level in November, immediately following the earthquake. This was true throughout much of the region, even in Santa Cruz County, which sustained significant physical damage.

There has been a great deal of discussion about the impact of the earthquake on the hotel/motel occupancy rates. Table 5-2 shows some impact from the event. San Francisco showed a 68 percent occupancy rate for rooms over the $110 price range for the first 6 months of 1989, through June. However, for the same period in 1990, the occupancy rate was 63 percent. If June 1990 is compared with June 1989, significant recovery in occupancy rates can be seen in the aftermath of the earthquake.

Therefore, there are basically two types of economic impacts. One is induced by a breakdown in the infrastructure. For example, the Bay Bridge was probably the single biggest factor that created temporary job and income losses. The second one is related to the economics of fear, which, for example, played a significant part in the decline in hotel/motel occupancy rates in the Bay Area. Tourism figures in the aggregate for the entire region indicate that there is no effect; but in San Francisco, there has been some decline resulting from the earthquake. Fear and anticipation of a recurrence of an earthquake has resulted in a decline of the tourism industry in the city.

TABLE 5-2 Economic Impact of the Loma Prieta Earthquake

Hotel-Motel Occupancy, San Francisco, 1989-1990

	June 1989	June 1990	Six Month June 1989	Six Month June 1990
Rooms over $110 rate	72.4%	76.0%	68.0%	63.0%
$80 - $110	82.0%	83.0%	80.4%	71.7%
Under $80	72.0%	80.0%	73.7%	67.4%
Average	73.0%	81.0%	72.0%	66.0%

SOURCE: Pannell, Kerr, Forster, San Francisco, July 1990

For example, eight major conventions were canceled in November 1989. Some have been rescheduled, but some have not. Also, fewer U.S. tourists are visiting the Bay Area; but there does not seem to be an impact in terms of tourism from the Far East or Europe. There is even an increase in tourism from Japan, perhaps because of good bargains and relative familiarity with living in seismic regions.

Much smaller regional areas must be examined in order to really assess economic impacts. For example, there appear to have been major economic consequences in Santa Cruz, the Marina District, and the Oakland area.

In the aftermath of the Loma Prieta earthquake, enormous traffic congestion occurred as the Bay Bridge went out of commission. Ridership on BART (Bay Area Rapid Transit) increased tremendously; but 6 months after the disaster, BART appears to have gained only about 2,000 permanent passengers. Immediately following the earthquake, BART ridership increased from 60,000 to 70,000 passengers per day.

A great deal of discretionary travel from East Bay to San Francisco appears to have stopped; people are staying home in East Bay. They are not crossing the Bay Bridge. The bridge seems to have become a psychological barrier.

The earthquake resulted in a significant decline in the supply of low-cost housing for the poor and for members of various ethnic groups because of the damaging structural impacts in the Oakland area, the area south of Mission Street in San Francisco, and the Tenderloin District (a region with a large Indochinese refugee population). The issue of housing for these impacted populations has become an important social and political concern in the Bay Area.

An interesting phenomenon may now be occurring with respect to real estate and land values. Although there has been a general decline in real estate values in the San Francisco area, the prices of land and buildings on bedrock have risen, while the value of areas involving higher seismic risk has declined in general.

On the whole, it is somewhat frustrating to try to assess the economic impact of an earthquake, particularly if large regional areas are examined. The various effects wash out. Part of this washout is explained by relief and rehabilitation money that has flowed into the region for repair and reconstruction. This inflow offsets some job losses and income losses.

With regard to the flow of funds into the Bay Area, only about half of what was initially expected has been received. This is a matter of much concern in the region. Often, initially there is a high level of optimism in the region about this "windfall" inflow of funds.

The relief money flowing into a region is not a benefit in a larger sense. Somebody is paying for it. The cost of this aid is being passed on to the next generation or to another region. People who live in safe areas subsidize those who live in dangerous areas. This is a matter of interregional transfer. But from a regional point of view, net transfers are good in that they minimize regional losses.

There are some important public policy issues that must be mentioned. One is that there has been a tremendous awareness of regional linkages in regional policy initiatives as a result of this earthquake. For example, we became very aware of transportation vulnerability issues in the Bay Area, particularly because of the damage to the Bay Bridge. This may have been a preview of transportation problems that the Bay area could experience, perhaps in 1998. But, as the memory of the disaster fades, some of these lessons and experiences are already being forgotten. The window of opportunity for instituting traffic-congestion-mitigation policies may already be closed.

In summary, if we look at the nine-county Bay Area, the economic impact has been minimal. If we look at the state as the basis for our analysis of economic impact, there is no longer even a blip. But there are certainly major economic impacts within the region, especially at the subregional and neighborhood levels. More attention needs to be paid to some of these issues also because good public policy must look at distributional effects, not just economic aggregate effects.

GENERAL DISCUSSION OF CHAPTER 5

QUESTION: It was interesting that you did not mention that there were few insurance dollars that flowed into the San Francisco area after Loma Prieta. There was quite a bit of money during the course of the recovery.

DR. MUNROE: That's true. It certainly helped the recovery, no question about it. The inflow of funds has created incomes and jobs in the region.

QUESTION: It appears that after the Loma Prieta earthquake, PG&E coped very well with repairing damage and restoring services. They did this by calling upon all of their personnel and also by using mutual aid from all around the western region from other utility companies. Now, this was a moderate event. What would be the special challenges that would be presented by a catastrophic event? What is the difference between the kinds of things that we have been looking at in terms of earthquake events and the major catastrophic events?

MR. EGUCHI: As I tried to indicate, it depends very much on the size of the event. In California we have a very special situation where there really are two major gas suppliers or distribution companies--PG&E and Southern California Gas. The sort of scenarios in California are really kind of independent events, so a lot of mutual aid resources and so forth can be expected to be to be there. Now, for situations like water companies, it is quite different. They are much smaller and, in terms of maintenance and supplies, it is much more limited. There would be a tremendous impact on those kinds of facilities.

DR. MUNROE: Let me add an example. PG&E had to shut off gas supplies to about 50,000 homes. Within about 2 or 3 days, we had to relight most of the pilot lights. We were able to do this with nearly 1,100 people from PG&E's and 350 people from Socal Gas from southern California. The latter went on our payroll for several days. If we are talking about a more significant earthquake, we could certainly obtain greater assistance from Socal Gas and elsewhere; we could also get assistance from Nevada and Oregon. The regional cooperation was absolutely first rate, and the inter-regional cooperation between utilities was impressive. We, as a society, do so well in times of crisis, it is just incredible.

QUESTION: But as Mr. Eguchi also pointed out in his presentation, when more and more are invited people in, that many more human resources must be managed. Where you are going to put these people; how you are going to house them, and so on?

MR. EGUCHI: In some of the smaller areas down south, there was a real problem, because there were no places to house the crews, and feeding them was very difficult. They had to go in and out of the region each day. Personnel management can be a significant problem.

DR. MUNROE: PG&E had a drill for a major earthquake, actually much more than a 7.1 earthquake; and that drill was extremely useful. I think preparedness and employee dedication was a key to our successful response, and I think it will have to be key to anything in the future.

QUESTION: Commenting on the previous question. I do not think you can extrapolate linearly. First of all, this was a remote earthquake. This was not an essential part of the Bay Area. The [PG&E] network was not damaged, its major facilities were not damaged. The distribution system in some localized places was lost because of the geology. A larger event, at the core of the Bay Area, would be qualitatively significantly different, not just taking more time to make repairs but actually rebuilding some facilities.

DR. MUNROE: Another factor was favorable for a fast recovery. For example, the Chevron refinery, one of the largest in the West, had no damage.

The water system was intact. Most of the utilities, including telecommunications, were really in very good shape. I guess this is what occurs in a moderate earthquake. The real impact on PG&E from the impact of the event was $100 million, and most of that was due to structural damage at two fossil-fired power plants and two or three major substations.

MR. EGUCHI: I think we had a very special situation during that event with regard to lifelines. The unique feature was that the epicenter occurred very much south of, say, the San Francisco Bay area, where you have a lot of urbanized systems and so forth. If you look at the areas where you had damage and where you did not have damage, there was a very strong distinction between those areas that were located on fill or were affected by some type of ground-failure amplification, which was actually very limited. If a large earthquake occurred closer to that area, the shaking effects would cause significant damage to facilities. Damage may not be so limited as what we saw in this event. I think we were very fortunate in that respect.

DR. MUNROE: Yes, there were many fortunate circumstances. The fact that a game of the World Series was being played that day may have saved many lives, because people left work early, before the earthquake occurred. Only 60 people died on the Cypress Overpass, and not 200 as initially estimated. The reason only 60 people were there instead of 260 is because all of those people were watching television at home and were not on that freeway.

6

The "Ripple Effect"

A great deal of concern has been raised about the economic consequences of a catastrophic earthquake extending beyond the region of impact to disrupt state, regional, or even the national economy. Of particular relevance is the concern raised about the ability of the insurance industry to cover losses from a catastrophic earthquake without having to sell off investments--stocks, bonds, real estate--an action that might have negative financial implications for regions outside of the disaster area. How farreaching could the economic impacts extend from such an event? What conditions are likely to impede or exacerbate this effect? What research exists to support the possibility of such an effect?

The two presentations in this session will explore the methodological and theoretical bases for projecting such an effect. **Barbara D. Stewart** has undergraduate degrees in economics and business administration from Beaver College. Ms. Stewart has worked for Stewart Economics, a consulting firm specializing in insurance issues. Her presentation will address the national economic impact of a catastrophic earthquake.

The second presentation is made by **Professor Leonard K. Cheng**, an associate professor of economics at the University of Florida. Dr. Cheng received a Ph.D. in economics from the University of California at Berkeley, where he specialized in international trade and trade policies under conditions of uncertainty. Dr. Cheng will assess the theoretical and empirical evidence related to economic "ripple effects."

PRESENTATION OF BARBARA D. STEWART

A catastrophic earthquake will have a national impact, and there will be national damage. The United States has a highly developed, specialized, interdependent, money economy. While those features make our economy productive and resilient, they also mean that an earthquake of the magnitude that we are contemplating in this forum will not be just a regional event.

There are three ways to view national economic damage: (1) disruptions to supply lines, (2) shocks to financial markets, and (3) drain on the insurance system.

There has been very little study of these consequences for obvious, very understandable reasons. It is quite human to focus on the human suffering and the physical damage that will occur immediately after an earthquake. The problem is that it is unknown, other than estimates of the physical damage, just how bad the general economic damage might be, and that uncertainty is a problem in itself.

The disruption in the earthquake area could easily break critical supply lines in the economy. It is very well known that gas and oil pipelines run through the New Madrid region and supply many businesses and individuals throughout the Northeast. We are also very critically aware that the important semiconductor industry is concentrated in California, and a catastrophic earthquake there would affect a wide array of other businesses, because suppliers there would shut down.

In general, shutting down most activity in the earthquake area will spread beyond that area as customers or suppliers are hit by the shutdown, and many businesses that are far away will suffer. There is a good chance that many of those, which might have been marginal to begin with, will just never start up again.

The inability to supply, the inability to sell, and the multiplier effects that will spread from the area are what many analysts are calling the ripple effect. Now, ripple connotes a less and less noticeable effect as the earthquake is spread over time and over space. But it is important for us to keep in mind that there are going to be more than just ripple effects, and they are not going to be orderly, spread over time and space. There are going to be some immediate and large impacts on the national economy, and those are going to come through the financial markets and through the insurance system.

The U.S. financial system is vulnerable to the physical damage of an earthquake. Just consider banks. Major banks in cities across the country are important switches in a complex financial network. They serve not only their regional economies, but are part of a national payments system. The money-center banks in particular transfer billions of dollars every day by wire around the country and around the world. One of the large California banks has estimated that if its central data processing were inoperable for 3 days, it would affect the entire state. If it were for 5 days, it would disrupt the whole U.S. economy, and if it were for 7 days, the world would feel it.

Having lived with the earthquake threat for a long time, California banks have some very sophisticated emergency planning systems, but whether those emergency plans contemplate everything in the area being shut down is something else. Another question is about banks outside of California that have not lived with this threat. Are they prepared for more than isolated events like fire or terrorist attacks?

Physical damage would also bring loan defaults. There have already been experiences with mortgage defaults after the San Fernando earthquake. Many people walked away from their homes if they had very little equity in them. The same thing is happening right now in the Northeast as property values fall and a lot of people walk away from their mortgages. It would happen again.

Commercial loan defaults would be even more serious, simply because the damage and the effects on other people would make it impossible for many businesses just to service their debt. The national question here is whether this country can deal with another round of bank failures, or call it another round of rescues.

How would our securities markets respond? I would not presume to predict what stock and bond markets would do. That is a folly for those who want to do it, but let us just think about it for a minute. Everything else being equal, if the assets that are underlying the securities that are being traded have been damaged, surely the prices of those assets would want to fall. But the greatest threat to security markets is not so much the damage but the uncertainty coming out of the damage. Financial markets work because of continuity and because of confidence. As we heard yesterday, our whole financial system is based on nonearthquakes. The United States economy does not have any experience, or in any other developed economy, with a catastrophic earthquake in a major business center. We just do not know what it means.

There are big questions of recovery. When? How? Where are the funds coming from? What else is going to be affected? What are the third-, fourth-, and fifth-order effects of this sort of thing? These are tremendous uncertainties, and if there is anything financial markets cannot stand, it is uncertainty. They can deal with good news, they can deal with bad news, but uncertainty is the worst. They are not going to be helped, either, by the fact that at the same time there are going to be tremendous demands for funds from the earthquake area in order to rebuild, and at the same time, the insurance industry is going to be dumping stocks and bonds on the market in order to raise cash to pay for claims.

The congestion in financial markets could have two effects. First, it would cause many businesses and governments to postpone financing for probably critical projects, for how long we just do not know. And second, it could have a liquidity effect on others who needed to sell securities to raise cash just for ongoing business needs.

When measuring systems are put together, specific things must be kept in mind. What is the cost of the things that are not done, the projects that are not built, the activities that are not undertaken?

After a catastrophic earthquake, what the insurance system would have to do to raise the money to pay claims and keep sound books of account would send another kind of aftershock throughout the country. It would happen in three ways in an increasing order of seriousness. First, as mentioned earlier, insurers would have to sell billions of dollars of securities to raise cash to pay claims. They just could not borrow. The needs would be too great and the debt service would be too much relative to the capital and earnings they have. These massive sales would depress prices in markets that already would be in turmoil because of the uncertainties over the earthquake.

Most affected would be the municipal bond market. Property and casualty insurance companies hold 20 percent of the municipal bonds outstanding in this country. In some years, they are enormous purchasers. They may take as much as 100 percent of the new supply coming on the market; in other years, they may not take much at all. But they are big players in the municipal market. Municipalities that wanted to raise funds, even if they were far away from the earthquake, may not be able to get into the market.

The second way the insurance system would disrupt the national economy would be from the magnitude of claims. Although the industry has more capital than the anticipated claims that we would get from a catastrophic earthquakes, that capital is not evenly spread throughout the industry, so we can expect a number of insolvencies to come out of the earthquake. The states have set up funds for dealing with the unpaid claims of insolvent insurers, but they typically do not pay in full, or else they pay over a very long period of time. And insolvent insurers would not just be those involved with earthquake claims. They would include some large, diversified national companies, so not just earthquake claims would not get paid. All over the country other kinds of claims (e.g., automobile, workers' compensation, and fire) would not get paid.

The third and most serious way that the insurance system would affect the national economy is what would happen after the insurance system had finished paying the claims. This is the real question. Earthquake claims would wipe out at least half of the industry's capital. Capital is the industry's ability to absorb and take on risk. Therefore there would be less capital, which would lead to shortages and higher prices of all kinds of insurance, regardless of location.

Why is this so serious? Well, insurance, like banking, is so interwoven in our everyday transactions that it is difficult to imagine an economy without it. We tend not to think about it, but it is very much integrated in everything we do. Insurance is an essential facilitating mechanism. By shifting and spreading risk, it lets an individual, business, or government pursue activities without fear of jeopardizing the household, the enterprise, or the institution.

How important it is for others that risk be shifted and spread is reflected in the fact that three-quarters of the property and casualty insurance sold in the United States is required by someone else. State statutes require automobile and workers' compensation coverage. Commercial and residential mortgage lenders require borrowers to have property damage and sometimes liability insurance.

Without insurance, many entities would have to bear their own risks or else curtail their activities. The economic and social costs, as well as the uncertainties, would be enormous. What would we do to replace insurance? Would we change our law? Would recovery in torts be permitted? Would we import it? There are tremendous uncertainties.

In conclusion, a catastrophic earthquake will do more than ripple through the economy. There is going to be an immediate impact on financial markets and the insurance system, both of which are national and international in scope. This is more than a measurement problem. It is also a conceptual problem. There are no models to go on. The United States today is not San Francisco in 1906. Nor are our major business centers Armenia, Iran, or Managua. We are dealing with a completely different order of magnitude, quantitatively and qualitatively.

There is a reluctance of participants in our financial markets to even discuss their vulnerability. Why should they undermine confidence if they do not have to? These uncertainties are useful to keep in mind when discussing the earthquake project and what to do with this kind of hazard, particularly when we get into issues of mitigation, because just as physical damage is not the whole story of catastrophic earthquake, mitigating physical damage is not the whole story either. It is going to be just as important to mitigate these uncertainties, that can have just as large an economic impact as the physical damage will.

PRESENTATION OF LEONARD K. CHENG

This presentation offers a few theoretical considerations which may be useful in assessing the economic ripple effects of a major earthquake (i.e., economic effects on areas other than the impacted region). In addition, some related empirical evidence will be provided to get an idea about the order of magnitude of these effects.

Since all regions of the United States are part of an integrated national economy, undoubtedly the economic effects of a major earthquake (taken to be in the order of magnitude 8 to 8.5 on the Richter scale) will not be limited to the impacted region alone. Hence, there will be ripple effects. The real question, however, is about their scope, intensity, and duration.

To see how the rest of the country will be affected by a major earthquake in the long run, imagine an artificial economy which is completely diversified in the sense that every household in the country (say, $\underline{n}$ of them) owns $1/\underline{n}$ of everything. In this fictitious world, if an earthquake hits, every household will be affected in the same way independent of the position of the epicenter. Suppose 5 percent of the country's total assets are destroyed; then every household will lose 5 percent of its wealth. If the loss in assets and people is permanent, then there will also be an additional loss in real income due to the reduced scope of specialization. If the earthquake causes a total economic loss of $50 billion, then this loss will be borne by all of the households in the country, which has a GNP of about $5 trillion per year and total wealth well in excess of $15 trillion. The impact per household outside of the impacted region is very small.

Obviously, the pattern of asset ownership is far from completely diversified. Assets in a particular region tend to be owned mostly by households residing in the same region. This implies that most of the long-run losses will be borne by the impacted region, and the losses to the rest of the country, perhaps in the order of billions of dollars, will be even smaller. To many they will be negligible. An important theoretical conclusion is that if the ripple effects are transmitted to a larger area through economic linkages, then the average effect per household outside of the impacted region will be smaller.

In the short run (anywhere from weeks to months, or even a few years), however, the effects outside of the impacted area are likely to be larger than the long-run effects, since adjusting to a changed environment takes time and can be costly. The severity of the short-run ripple effects depends on the degree of substitutability in the economy. In one extreme, if (1) the products produced in the impacted region are vital to production or consumption in the rest of the country, and (2) substitutes are not available or producible outside of the impacted region (not even in foreign countries), then the short-run ripple effects would be disastrous. In the other extreme, if there is a great deal of substitutability in both production and consumption, then the short-run effects would be approximately equal to the long-run effects.

As an example, if a certain computer chip produced in the Silicon Valley became unavailable after a major earthquake hit California, it can be substituted with identical or similar chips produced in other parts of the United States or in foreign countries. Even if substitutable chips are not available (which is extremely unlikely), so that products, incorporating the chip cannot be produced, consumers may still substitute these products with other products because the same need can be met in different ways. Finally, consumers can also substitute current consumption with consumption in the future, when the chip becomes available again.

An advanced economy like the United States is not only highly integrated (which may give rise to enormous adjustment costs under extremely adverse conditions), but also exhibits a great deal of flexibility and substitutability. As a result of this second characteristic, it is unlikely that the short-run ripple effects would be much higher than the long-run effects, which would be very small in relative terms. For instance, there is no evidence that the San Francisco earthquake on October 17, 1989, has had much of an economic impact outside of the Bay Area.

Without the economic linkages, the different regions of the country would be like isolated island economies. A region which is hit by an earthquake, large or small, will be the only party to bear the *entire* burden. In contrast, an advanced and highly integrated economy allows risks to be shared through trade and credit relations as well as ownership diversification. It is analogous to numerous small nets knitted into a big net. When an object falls on any small net, its impact will be partly borne by other nets. To ignore the benefits provided by the economic linkages, including a greater capacity to deal with disasters, and to view them primarily as the nodes of a network by which diseases are spread would be theoretically wrong and very misleading.

On the one hand, the loss of capital and equipment due to an earthquake implies the loss of income in the impacted region derived from the original economic activities. On the other hand, the need for reconstruction would increase employment. Provided that there are enough past savings, credit, insurance payments, and investment by outside investors, it is likely that the level of economic activities in the impacted region could quickly exceed that before the earthquake. It is also very likely that producers outside the impacted region would benefit from the reconstruction effort in the region. Of course, any surge in economic activities for the purpose of replacement and repair will taper off once recovery is more or less completed. These predictions about the economic effects of a major earthquake on the impacted area are consistent with simulation results.[49]

The economic data presented by Dr. Tapan Munroe in connection with the recent San Francisco earthquake have revealed the remarkable resiliency of the impacted region's economy. The data suggest that, at the county level, the quake has had little economic effect. To find drastic, localized effects which are marred by aggregates, one has to look at data for cities or even blocks within cities.

Full and just insurance payments to the insured victims of earthquakes will speed up recovery in the impacted region. An effort by the insurance industry to do its best to settle claims in full and speedily not only is *not* a disruptive force to the national economy, it would lead to good business for the insurance industry in the future.

If the effect on future demand for insurance, is ignored then any shortfall in insurance payment for validated claims amounts to a transfer of wealth from the insured victims to the shareholders of the insurance companies. If the insured victims are able to undertake the same needed repair and replacement without the full payment from the insurance industry, and if their marginal propensity to consume out of wealth is identical to that of the insurance companies' shareholders, then after an earthquake hits, how much the insurance industry ends up paying will only affect the wealth distribution of these two groups. However, since how much and how fast the insured victims can replace and repair their damaged properties are likely to depend on the payments they receive from the insurance companies, and provided that the earthquake victims' marginal propensity to consume should be no less than that of the insurance industry's shareholders, we can conclude that full (and justified) payments to the insured victims will lead to a faster recovery in the impacted region and greater output and income for the country as a whole.

The effect on the financial markets of the insurance industry's need to liquidate some of their bonds to settle claims (say $30 billion) would be minimal because of the size of the securities market. The outstanding Treasury bills and bonds alone are in the order of trillions of dollars, and recently the federal government has been borrowing hundreds of billions of dollars each year. For similar reasons, a major earthquake's destruction will have a very small effect on the stock market. For example, even though the San Francisco earthquake last October inflicted about $8 billion of damages to man-made facilities (which includes $2 billion of damages to the transportation system), it did not have much of an adverse effect on the stock market. Indeed, the stock market has since been achieving record highs.

To get an idea about how the insurance industry may be affected by a major earthquake, we can look at the performance of the industry in 1989, when it was hit by several catastrophes. According to the insurance industry's trade journal, Best Review,[50] insured losses caused by (1) Hurricane Hugo, (2) the San Francisco earthquake, and (3) storms, floods, tornadoes, and the

Phillips Petroleum plant explosion were $4.1 billion, $1.1 billion, and $1.8 billion, respectively. Despite these major catastrophes, insurance stocks performed favorably, especially in the area of property/casualty. Stock values surged to their highest record levels. The overall A.M. Best Insurance Industry Stock Index went up by 37 percent in 1989, compared with a 25 percent growth of both the Dow-Jones Industrial Average and S&P 500.[50] Moreover, the ability of the insurance industry to provide service to the economy as a whole is not limited by their current capital and surplus. Since capital is mobile across industries and will move to whichever industries are profitable, the capital base of the insurance industry is not a constant equal to their current capital less insured losses. Instead, in the long run it will be determined by demand for insurance, including insurance against earthquakes.

Reinsurance is a device for spreading risks among the insurance companies, including foreign companies. Like all economic linkages discussed above, it spreads the burden of shocks to a large number of participants, and therefore is definitely *not* a cause of disruption to the industry. Furthermore, participation by foreign companies in insurance and reinsurance has significantly helped the U.S. insurance industry deal with the insured losses.[51]

On the basis of the theoretical considerations and available evidence from the San Francisco earthquake last October, it can be concluded that the economic ripple effects of a major earthquake would likely be small. Moreover, due to the scope of substitution in production, consumption, and investment, any significant ripple effect will be dampened quite quickly. Of course, rigorous research is needed to obtain precise estimates of the magnitude and duration of the ripple effects under different scenarios.

One useful approach is simulation, with a carefully constructed model of a regional economy embedded in a national economy which incorporates real-world economic linkages between them. For example, an extension of the regional econometric model constructed by Ellson, Milliman, and Roberts, which incorporated supply-side constraints and spatial disaggregation, would be a fruitful avenue. Since the linkages between the impacted region and the rest of the country need to be modeled in detail (such as spatial disaggregation), the level of disaggregation within the impacted region can be reduced accordingly to retain tractability.

A report prepared by Japan's Tokai Bank attempted to estimate quantitatively the economic ripple effects of a major earthquake in the Tokyo area on the rest of the world. Unfortunately, the estimates were derived from highly questionable assumptions. For example, it assumed that the earthquake in Japan would raise the U.S. Treasury bill rate by 5 percent. Given that Japan's total lending to the United States is only a trivial portion of the U.S. capital market, it is implausible for the real interest rate to rise by such a great magnitude. Since the world capital market is highly integrated, an appropriate comparison is the total reduction in lending by Japan due to the earthquake against the size of the world capital market. From this perspective, the effect on the interest rate in the United States must be quite small.[52] My colleague, Dr. David Denslow, pointed out that, with the beginning of

World War I in July 1914, long-term corporate bond yields in the United States increased from the prewar (June 1914) rate of 4.06 to 4.22 percent in December of the same year. By January 1916 they were back down to 4.06 percent.

The methodologies for estimating the economic ripple effects of a major earthquake in the United States exist, but care must be exercised in constructing an appropriate model which can be used to generate meaningful and reasonable results.

Acknowledgment is given to Dr. Jerry Milliman for helpful guidance and support.

GENERAL DISCUSSION OF CHAPTER 6

QUESTION: Dr. Cheng, you said that there were no effects on the economy from many disasters, or the economic effects were minimal. What level of a disaster, what level of a catastrophe, would cause effects on the insurance industry? Loma Prieta did not. What is the catastrophic disaster that would trigger some effects?

DR. CHENG: Of course, if the quake gets bigger, it is going to have more of an effect. That is obvious. But how big is the big one that we are going to prepare for? Are we talking about magnitude 10 or 9.5 on the Richter scale? And I am sure that if the quake is big enough, available surpluses could be wiped out and, therefore, in the short run there would be a serious effect. But still I do not believe that is what we are talking about here. We are talking about the effect which is great but not of magnitude 9.5. In the Blue Book published by the Earthquake Project, they estimated a $50 billion loss.[53] Now, of course, if it is $5 trillion, then I am sure it is going to have a terrible effect. But is that likely to be from an earthquake?

I think we had better talk to the loss-estimation experts. How likely we are going to suffer a $500 billion loss from one single earthquake?

QUESTION: I thought I heard one of you say that when a great earthquake happens, it is going to be really big from an economic impact point of view, and another one of you say it is going to be really small. What is it really going to be?

DR. CHENG: The answer lies in the following. The key to the different predictions lies in the degree of substitutability. Do you believe that this economy has no scope for substitution? If that is what you believe, and if you feel that in the real world, everything produced in every corner of the world is vital to our lives and without it we die, then there is no way you can substitute. If that is the case, then you get the first scenario--that everything will be terrible. I do not believe that is the world we live in.

In the world we live in, we have a great deal of substitutability and a great deal of flexibility. Therefore, I tend to believe on a theoretical basis and from empirical observation that the effect will be very small outside the impacted region.

MS. STEWART: I certainly agree that we have a very resilient economy, and it is possible to substitute. It is important to keep in mind the short-run dislocations and the long-run readjustments. In time, definitely we will adjust. We will find substitutes, we will arrange new sources of supply. Our financial markets will recover. Both Germany and Japan rose out of the ashes of World War II.

The short term is what counts. It is important to look at the short-term dislocations, where we disagree as to the magnitude of some of the problems.

Part of this disagreement may be related to this one issue that is really not emphasized much in the Blue Book--that is, the uncertainty that is going to be facing many parts of our economy. This is something that is very difficult to deal with on a theoretical basis. We can model all we want to and crunch the numbers all we want to, but this is going to be a tough thing to deal with because of the lack of prior experience. It is something we should advisedly consider and not ignore as we get on with our abstract models and our measuring and other devices to deal with trying to judge the magnitude of this thing.

DR. CHENG: I would like to add one point. I am not denying at all that there will be serious economic consequences. What I am saying is that most of that will be in the impacted region. If you are going to look for devices to address the problem, I think that is where you look, not from the national economy, given the usual sizes of the effect of an earthquake that we are talking about.

QUESTION: The All Industry Research Advisory Council was interested in the ripple effect on the insurance, reinsurance industry for a large, catastrophic event. We chose to look at the simulated effect of a $7 billion insured hurricane on the Gulf or East coast, and then a combination of two of those events in the same year. We simulated the effect, looking at the various possibilities where you have clusters of properties along the Gulf Coast--Houston, Miami--and then made estimates from each insurance company where their losses were reinsured. We tried to follow those dollars--the ripple effect--to Europe and back to the United States.

The insurance industry can handle a $7 billion loss without much problem. When you get up into $14 billion, you get into a marginal area. If you get much above $14 billion, there is a large internal ripple effect, which is much larger than was implied in your model. Could you comment on that?

DR. CHENG: I think you know better than I do about the internal workings of the insurance industry. But if you are talking about the need to raise even $30 billion from the security markets, that is nothing compared with the overall size of the security markets. Let us not lose perspective on that. We are talking about a huge capital market out there. I am not denying that the insurance industry would have more difficulty when it has to make more payments. That is to be expected. But I believe that is an industry-specific effect. Going back to the ripple effect, I think it is negligible.

QUESTION: I would like to refer to an earlier point regarding the difficulty of assessing economic impacts, even on a nine-county basis, even on a single-county basis. The resiliency, the substitutability, and the flexibility of the economy of the San Francisco Bay Area was really, really impressive. It is not just silicon chips we are talking about. We are talking about petroleum refining, smokestack industries, about a very diversified economy.

I know of five plants that were out of commission--four of them got back into production in 3 days. One of them was totally ruined, but they found a new facility within 2 weeks and got back on their feet. This is something we have to remember. The evidence that we have so far on earthquakes in this country would suggest that the damages, or rather the losses, are certainly on the lower end of the spectrum rather than on the higher end. So far, the conclusions that have been made about overwhelming losses from catastrophic earthquakes suggest that these losses may not be all that great.

Now, we always go back to 1906; but the U.S. economy is not on a 1906 basis in terms of infrastructure, in terms of institutions, in terms of technology. It is absolutely incorrect to extrapolate damages and losses on the basis of 1906, 80 years hence. Since it is unrealistic and incorrect, I would suggest that we try to run simulations on other data. For example, in a study done by Jerry Milliman in 1983, a county-by-county model of the California economy was used. A catastrophic earthquake did not result in catastrophic losses for the state. We need to be very careful in making basic assumptions, because that is what makes all the difference.

I am really concerned about the highly pessimistic nature of the base case that I have heard.

QUESTION: There has been a prediction, I am sure you know, that the first of December we are going to have a major event in the New Madrid area, so probably we may learn whether we are right in our estimates or not. The insurance industry came up with an estimate of insured losses--something in the area of $50 billion. Those are the insured losses, but insured losses are only a small fraction of a total loss to the economy. Public properties, for example, are not insured. If you add those, a $300 billion top estimate is necessarily wrong.

The second point I would like to make is that being in the insurance industry, we have to provide for, if not the worst possible case, fairly near it. If anything can happen, it will. I want to look at the liquidation of securities. For technical reasons, the property and casualty insurance industry is very largely invested in municipal bonds, because they receive favorable tax treatment. In any event, the insurance industry is so invested and it is not a liquid market. We are large purchasers of municipal bonds, and we know that the market could not handle an order to liquidate $30 billion of municipal bonds. Because this could not be handled, we would be obliged to liquidate stocks. Even though the stock market is a very wide market, I think that the specialized effect of a demand by the industry to raise $30 billion would have a very material effect on the security markets. That is a short-term effect, obviously.

In the long term, the market is huge. It will absorb it, but if we wanted to liquidate it overnight, relatively speaking we would have a problem.

DR. CHENG: First of all, the way to handle the prospect of huge losses is to accumulate surpluses. I know very little about the insurance industry, but I think what the insurance industry can perhaps do is to ask the government to change the law so that it can accumulate surpluses before tax.

My second comment is related to your suggestion about the liquidity of the municipal bond market. All bonds, to different degrees, are substitutable. If the yield on the municipal bonds is very high because of a "fire sale," then a lot of people will sell other bonds and buy municipal bonds. All bonds, while not perfect substitutes, are close substitutes in the eyes of the investors. I am not trying to deny that there will be difficulties in the short run. But I believe that the ripple effects will be minimal.

QUESTION: We are not allowed to accumulate any reserve tax-free, whatsoever, for an event that has not yet happened. If we go a year without an earthquake, the IRS considers that we have made a profit on whatever premiums we have collected during that year, and taxes it as such. It is not very easy to build a reserve for future events. Now, of course, we do build surpluses to the extent we are allowed to by various insurance commissioners around the country. But no doubt that a carefully constructed provision allowing insurers to accumulate reserves specifically earmarked for a major event, not necessarily only an earthquake, would be a very helpful thing. For example, in Japan, we are allowed to accumulate such a reserve. It gives you a sense of comfort.

However, the industry's chances of getting such a change through the Congress at this time are very limited. We would welcome anybody to speak up very loudly and clearly outside of our own industry that this is a desirable thing to be done, and it could be, provided that be totally invested in United States securities.

DR. CHENG: The current rules applied by the IRS only work if the annual expenses and payments reflect the expected payments for events that recur on an annual basis. But when you are talking about something that might not occur for 100 years, the actual payment in one year would bear little relationship to what you expect to pay in the future.

QUESTION: When we do have this major 8.3 event, it will likely take place within the context of increased overall seismic activity in that area. There will be a number of earthquakes, not just one. An event of that magnitude will probably be followed by some very large aftershocks in the ensuing months, and maybe even years. For example, the New Madrid earthquake was not really a single earthquake but a series of events, including three very large ones. Would this likely circumstance change your predictions or your thinking about the economic impact? How do you think the markets will react to this?

DR. CHENG: Well, if the aftershocks come immediately after the big quake, then the reactions would be as though they all occurred at the same time. If the aftershocks come only after a long period of time, that will give room for the market and people to adjust and revise their expectations, therefore revising their preventive measures accordingly.

MS. STEWART: I would tend to disagree with that. I think if you have a second a year after the first one, it would have a very serious effect, that it would certainly question any efforts to rebuild in the area. Individuals who lived in the area and businesses that were there would question whether it made any sense to stay. There would be some real decision-making going on as to whether they should pick up and go somewhere else.

The same thing would hold true for the reaction of some of the financial markets. After they had absorbed the first shock and gotten back to working more or less normally again, the second one--although it would not have the impact of the first--would certainly undermine a lot of the normality that had been reestablished in the markets.

It is a very good question, and one worth thinking about. It also raises the earlier question about how much capital the insurance industry is going to have left. This is the real question. I do not think the industry is worried about paying that first round of losses. What happens with the next round and the next round and the next round? Capital does not move that quickly between industries and sectors in the economy. Certainly in the insurance industry, the way it has been losing money on its own without any catastrophes, it has not been able to attract new capital. I imagine it could attract capital but at a very, very high price and perhaps in limited amounts, so that we would also be talking about very large increases in insurance premiums and perhaps in all kinds of insurance premiums.

In other words, there are going to be effects upon effects upon effects. We are dealing with a very complex problem. These things will adjust over time, that is true. But in the short term, there are going to be some very serious problems.

QUESTION: I have a couple of short comments. First, you know, economics is often referred to as the "dismal science." It is good to have a cheerful economist like Dr. Cheng. I think there are some conceptual problems here in the way that we are looking at this problem. The first conceptual problem is the loaded language. For example, the term "catastrophic" earthquake. This term begs the question of what a very large earthquake will do. The real concern is, we need some very cool and careful studies of the impact of certain kinds of very, very large earthquakes--economically, physically, socially, and politically. A catastrophic earthquake has been defined as one which disrupts the national economy, etc., etc., etc. From the very beginning that definition set the wrong conceptual tone.

The second problem concerns the question of uncertainty. If you look forward on a short-term basis, there is great uncertainty. If you look back on a long-term basis, there is no uncertainty. If you look back, for instance, at the history of cities--which I am interested in doing--you find that cities have survived for hundreds and thousands of years in highly earthquake-prone areas. The occasional earthquake is, in effect, a natural phenomenon which is accommodated by the city, and the city continues because the reasons for its continuing are much, much stronger than the effect of the earthquake. So what do we really mean by short-term/long-term?

As you look forward, I would suggest there is much less uncertainty on the long-term basis than on the short-term basis. We are a very short term society, and this affects us in many ways other than earthquakes. You know, we have a 1-year business orientation to the bottom line. We have a 4-year political orientation to what is going to happen while in an office; this cuts against the way in which nature works, which is on a very long term basis. We have got to adjust our social and economic systems to the pattern which exists and to taking care of ourselves on a long-term basis, whatever that may be.

For you, 1 year may be short-term. For me, that may be long-term. Fifty years may be long-term. For nature or geology, that is a gnat's eyebrow.

QUESTION: Dr. Cheng, in your closing remarks, you mentioned that the Tokai Bank estimates it were based on highly questionable assumptions. Also in your closing remarks, you mentioned that methodologies do exist which can be used to make the precise estimates that you feel are needed. Were the assumptions the only problem with the Tokai Bank study? Was there a solid methodology used there or not?

DR. CHENG: I only read the abstract, and I really have not seen the details of the methodology. But given the fact that some regional economic models have been developed, the next step is to put them in the context of a national economy model and build in realistic economic linkages, that is, linkages that come from the real world as opposed to what we assume.

Now, talking about the Tokai study, it makes several unrealistic and wrong assumptions. For instance, it assumes that the impact would hurt the United States. That would be opposite to my prediction. If Tokyo were wiped out, the U.S. producers are in good shape; we are going to supply stuff to the world market because they are not coming from Japan. We are also going to supply goods to Japan because they need them for reconstruction.

My specialty is international trade, which is concerned about interdependence all the time. The Tokai study was an attempt to come up with quantitative estimates, but I think it failed because it made very unrealistic and wrong assumptions.

MS. STEWART: I do not think any of us would disagree that reconstruction after a catastrophe, be it an earthquake or a war, would be stimulating to the economies that would help in the reconstruction, just as we benefitted after World War II. This forum focuses on what to do to relieve the damage that is going to occur in the devastated area, rather than who is going to benefit from someone else's loss. I do not think there is any disagreement as to the stimulative effect of reconstruction.

QUESTION: Is there an analogous case that may lead to some testing of your models? For example, the reunification of East and West Germany may have an equivalent effect of a great earthquake. They have to rebuild the infrastructure, and I understand that estimates are something like $60 billion a year for the next 5 years, which is equivalent to $300 billion.

DR. CHENG: I think that is a very good point, and I would like to relate that to what Professor Hal Cochrane presented in his summary. There are many big events, and earthquakes are by no means the largest events in terms of economic losses. As you know, the German reunification would require hundreds of billions of dollars to reinvigorate East Germany. That, of course, is going to have an impact in terms of the interest rate in the same way that the Tokyo earthquake will have. That will drive the interest rate up, and resources will be diverted away from other borrowing countries to East Germany. That is a very good case, and a great test of what we predict about these ripple effects. If that is the only event, I do not think that will affect the capital market very much. But if you combine that with, let us say, the economic reform in the Soviet Union, you may be talking about several trillion dollars.

Let us also not lose sight of the fact that right now, the less-developed countries already owe $1 trillion. When comparing $50, $60 billion with the overall size of the world capital market, of which $1 trillion is only a small portion, that is not going to have any major effect. Like the figures that I gave about World War I, the impact on the interest rates is a very good indirect test of the predictions. The effects will be small unless the events all occur at the same time.

QUESTION: Apparently, the markets at this point seem to agree with that; the markets are remaining stable.

7

How do Current Relief Policies Affect Recovery Efforts?

A variety of state and federal policies and programs exist to assist both the public and private sectors in postearthquake recovery efforts. The purpose of the next three presentations is to describe what policies are currently available to assist in economic recovery and to identify the limitations of those policies if a catastrophic earthquake were to occur. Special emphasis will be placed on the consequences of these limitations outside of the direct impact area.

The first presenter is **L. Thomas Tobin,** the executive director of the California Seismic Safety Commission. He is a registered professional engineer in California with a graduate degree in geotechnical engineering. Mr. Tobin will provide a state perspective of economic-recovery issues.

Robert G. Chappell will present a federal perspective of earthquake-recovery policies and programs. Mr. Chappell, a federal employee since 1958, is a member of the Senior Executive Service. He is currently the assistant associate director for disaster-assistance programs at FEMA, which makes him responsible for managing the federal response to presidentially declared disasters. His topic covers the federal disaster-relief programs following a catastrophic disaster.

Richard J. Roth, Jr. has a bachelor's degree in engineering, a master's degree in economics and statistics from Stanford University, and a law degree. Since 1984, he has been the assistant insurance commissioner and chief property-casualty actuary in the California Department of Insurance. His presentation will cover the role of insurance in economic recovery.

PRESENTATION OF L. THOMAS TOBIN

The threat of earthquakes in California is real, but it is not magnitude 9 earthquakes and it is not Armageddon. Questions about whether we should rebuild or not are not legitimate. Questions regarding whether California cities will die are not realistic. Even after the worst earthquake imaginable, we will rebuild.

If anything comes out of this forum, I hope it is a better dialogue among our different disciplines. There is a real need for us to come together. If I were a member of Congress, after hearing the different views expressed, I would not even consent to a hearing on federal participation in the matter of earthquake insurance, let alone vote for it. Yet, I believe federal participation in earthquake insurance is very important and necessary, and it should be sooner, not later.

I live in the Bay Area, and am quite happy and comfortable living there. I expect a major earthquake to occur. I expect to repair the damage to my house, and help my neighbors, and go back to work in a few days. Our experience is that after a major earthquake, which may be catastrophic in terms of its economic effects, 80 percent of the buildings will still be standing and in use. One problem will be: how do we get people to work the day after the earthquake? I am not trying to downplay the deaths, injuries, and damage, but only to make the point that more will be standing than will be lost. Economic recovery will begin with a base of existing, functional buildings and structures.

That concept has been lost in much of what has been presented. Our model and scenarios must be legitimate. Clearly, more information is needed.

The State as a Financial Partner

Although the state is part of the recovery process as a supplier of capital, it has limited resources. The California Legislature certainly demonstrated the fact in July 1990 by taking 43 days beyond the constitutional deadline to pass a budget. State government is in dire straits economically, in spite of a healthy economy.

In addition, California lacks flexibility to spend tax revenues. The General Fund, the primary source of state recovery-financing money, comes primarily from taxes. State expenditures are limited by the Constitution to annual increases no greater than a combination of population growth and the consumer price index, and experience shows that state programs grow faster than the combination of those two elements, so the competition for General Fund monies gets more intense each year. The problem will get worse before it gets better.

Constitutional and statutory language limits the amount of money California can raise and spend and how it can be spent. Over 82 percent of the General Fund expenditures are fixed by law and the Constitution. For example, the Constitution guarantees a minimum of 40 percent of the General Fund for K-through-12 education, even when we have an earthquake. While that 40 percent can be suspended for 1 year, it cannot be reduced in subsequent years without a constitutional amendment, even if there is a fiscal necessity to do so that was precipitated by a natural disaster, such as an earthquake.

Other programs that are certainly not frivolous are guaranteed cost-of-living adjustments by statute. The governor and the Legislature cannot change another 42 percent of the expenditures without passing new and controversial laws to reduce support for health and welfare, higher education, prisons, property-tax relief, and other critical programs.

The only areas where there is budget flexibility are in natural resources, state consumer services, business, transportation, housing, and a few other programs. Thus the state, despite a $54 billion budget, lacks flexibility and the capacity to redirect funds after an earthquake.

California also lacks flexibility in raising new revenues. Even if the spending limitations were circumvented, the Constitution limits increases in state and local government taxes. After the Loma Prieta earthquake, the Legislature and governor agreed on a 1/2 percent increase in the sales tax for 13 months, but there is little political support for increasing personal income tax, sales tax, bank and corporation tax, or insurance premium taxes, except for a short period. Support might exist for increasing the sin taxes--tobacco, liquor, and horse racing--but those only account for 1 percent of the General Fund. There is not much there.

General obligation bonds are thought to be the silver bullet. However, they must be sold after the disaster--at a time when the bond market may be unhealthy or unreceptive to California obligations. In any event, the principal and interest must be paid back out of the General Fund. Raising money through general obligation bonds is slow and uncertain. In California, they require approval by the voters, and a special election may be needed. The state of California has limited resources.

While the state still is struggling with repair of state buildings and coping with disrupted operations after a disastrous earthquake, ongoing demands on the General Fund will not go away. Citizens in the rest of the state outside of the damaged area, and those who reside and work in the 80 percent of the buildings inside the disaster area that are still serviceable, need normal state services to continue without interruption.

Disasters in California have been *affordable*. California lacks experience in recovering from a major urban earthquake. There is no model. Damage in the Coalinga earthquake was about $35 million, in the Whittier earthquake about $350 million, and in the Loma Prieta about $5.6 billion. Prior to Loma Prieta, California had nine presidentially declared natural disasters with total losses of less than $2 billion. We have a lot of experience with disasters in California, but they have all been relatively small. Our government and financial systems have not been tested.

Local governments in California have similar limitations. Very few have large funds for economic uncertainties. There are constitutional restrictions on increasing the property tax or sales tax without voter approval. Other funds, such as from fees and licenses, can be raised, but must be used for increasing the services. Increased income from building-permit fees only covers increased costs for plan checking and construction inspections. In fact, local government in the damaged area tends to lose revenue from sales tax and property taxes.

Federal relief funds are not a sure thing. Every year, FEMA receives an appropriation for the amount of disaster aid believed necessary. When that is exceeded, it takes congressional action to increase the money in that fund. After the Loma Prieta earthquake, California sent a delegation of California legislators to Washington to seek federal legislation to increase the monies and to provide money through the Department of Transportation for rebuilding our transportation system. Approximately $3.45 billion was eventually appropriated, but the message they brought back was that Congress was extremely reluctant to appropriate the funds for use in California because of federal budget problems and because of the perception that California's economy and government should be capable of financing recovery. We cannot depend on the federal government to automatically provide an unlimited amount of recovery funds.

State and local governments are victims and will suffer a large share of the losses to government facilities. The Loma Prieta damage was estimated to be $5.6 billion; of that, $2.3 billion--40 percent--was for damage to public properties. State and local governments are generally uninsured, although the state also had some insurance covering lost toll revenues. We must assume that government losses in any major urban earthquake are substantial, and that rebuilding damaged facilities will be a high priority.

Sufficient state aid is not a sure thing, and this can be illustrated by the way California coped with the Loma Prieta earthquake. After it occurred, the governor first drew on his Fund for Economic Uncertainties. At the beginning of the fiscal year in July 1989, that fund was probably around $800 million, but the earthquake was not the only uncertainty to occur during the year. After the earthquake, the governor called an extraordinary session of the Legislature. Although an extraordinary session may not have been necessary during either the response or the initial stages of the recovery phases, politically it was necessary for state government to demonstrate concern and the ability to respond. The Legislature passed 12 identical bills in the Assembly and in the Senate simultaneously. All 24 were signed by the governor. The most important bill (ABx 48, Assemblyman Areias; and SBx 33, Senator Mello) that passed during the extraordinary session increased the sales tax by a quarter cent for 13 months. It filled an obvious need, and it appeared to be a small sacrifice for Californians to make after the earthquake. But passage was not automatic. The debate was heated, and the vote was close. Passage of a revenue bill takes a two-thirds vote in each house of the Legislature. The measure passed the Assembly with a 60-17 vote, even though there was near-unanimous agreement that the state should pay a share of recovery. Even after a disaster, there remained strong philosophical opposition to increasing taxes. Nearly one-third of the Assembly believed that the way to cope with the need for revenue was to cut other state programs. Even from the state of California itself, a large amount of disaster aid is not a sure thing.

About 6 months after the earthquake, state spending was reduced by cancelling outstanding contracts and freezing new equipment purchases and some hiring. But since the damage was so limited, the rest of the state was unaffected, and the demand for ongoing state programs continued. The business of government has to go on.

The temporary quarter-cent sales-tax increase was estimated to raise about $800 million; however, these projections are turning out to be high. Receipts after 7 months of the 13-month period are about $354 million. It appears that it will raise about $700 million over that 13 month period. A total of about $1.5 billion will be raised from the toll-interruption insurance, the Fund for Economic Uncertainties, and the quarter-cent temporary sales tax. Since then, some general obligation bonds also were approved to repair four damaged state buildings.

Cash flow must be considered. Although we are raising less revenue than projected, our recovery expenditures after the Loma Prieta earthquake have not caused a cash-flow problem. Nevertheless, the amount of money needed, and when it is needed, affect the ways the funds can be raised and the earthquake's economic impact. The speed with which holdings in various markets are liquidated affects their value, and the time taken to raise revenues from other sources affects the tax rate and political acceptability. Timing of recovery funding needs more consideration.

Paying for the Loma Prieta Earthquake

California has authorized bonds to finance some repairs to damage from the earthquake. State Proposition 122 authorizes a $300 million bond measure for retrofit of state and local government buildings; approximately $150 million will be used for repairs to three state office buildings. California voters also passed Propositions 108 and 111 to authorize transportation bonds and increase gas tax for new freeways. A portion of these funds will be used for repairing damage and retrofitting existing structures. Public support for earthquake safety was perceived as so strong that supporters aired TV spots that stressed earthquake retrofit rather than traffic congestion.

According to the State Department of Finance, the state's expenditures and obligations for the Loma Prieta earthquake (that is, commitments that have been made) to date are about $120 million. This compares with the $354 million income from the increased sales tax to date. The same sources indicate that the federal government has obligated $1,489 million. By comparison, the State Disaster Field Office indicates that the state's expenses to date are $340 million (about three times as much as the Department of Finance estimates as spent or obligated), and that the federal expenses are $1,500 million.

Ten months after the earthquake, its cost to the state and the federal government is still uncertain, but the numbers clearly indicate that rebuilding capital is not all needed immediately, and after 7 months (the date of my figures), less than half of the amount needed has been spent.

Observations on Economic Impact Studies

Reasonable earthquake events should be used to provide a realistic view of the range of losses. In California, studies should be based on magnitude 7 events occurring in urban areas. That is the size of event expected on the Newport-Inglewood fault, or the Elysian Park thrust fault underlying the Los Angeles basin, and the events expected on the Hayward fault, the Rogers Creek fault, and the San Andreas fault in the Bay Area. There is about a .9

probability that one or more of these earthquakes will occur in California within a 30-year period.

The probability of a magnitude 8-or-greater earthquake occurring is quite small compared with the more frequent magnitude 7s. It makes far more sense for us to be planning for two magnitude 7s within a given period of time, rather than a magnitude 8.

We need more loss studies. Old numbers are being used, even though better methodology and data are available. The numbers that we use in California come from a FEMA-USGS-NSC study in the late 1970s. It is a shame that 10 years later we still must pursue major public policy initiatives with old numbers based on out-of-date methodology. We need new loss studies.

Even though I do not disagree with the Earthquake Project's numbers, and even though I use them, I am concerned that the 40 percent of their estimated losses, those for losses due to fire following earthquake and workers' compensation, to my knowledge are based on one study on each topic. Although I do not have any problems with either study, and in fact respect the authors, I am surprised that an important public policy initiative would go forward based on only two studies. To maintain our credibility, we need additional points of view. We need additional loss studies and additional dialogue aimed at building consensus on the losses we can reasonably expect.

We need a consensus. When experts disagree and legislators have doubts, nothing happens. The need for federal participation in a catastrophic-earthquake- insurance program is so important to the public, the state of California, and the insurance industry that it is imperative we reach an understanding.

Earthquake insurance and disaster aid must be evaluated as part of an economic-recovery-management program. At present, the economic aspects of recovering from a major disaster are not understood. We do not know how to speed the recovery; we do not know what are efficient investments; we do not understand the time dimension. Our state and local disaster aid programs are largely humanitarian aid and are not tailored to assist economic or business recovery from an earthquake. Because the amount of recovery money available may be quite limited, we have to know how to best spend what we have. Certainly the wise and timely use, not political use, of recovery dollars determines the extent of ripple effects in the economy.

A simple model relating direct losses, indirect losses, and recovery time illustrates our problems and suggests policy needs. Direct losses mostly occur the moment the earthquake occurs, while indirect losses continue to increase until recovery is complete. Management of indirect losses is of foremost importance, since they can be double or triple the direct losses.

Indirect losses vary with a number of factors, but the most important variables are the amount of direct loss and recovery time. If recovery time is shortened, indirect losses are lessened.

The second variable is the amount of physical losses. Indirect losses and the length of time to recover increase as the physical losses increase. If we lessen the physical losses through mitigation, recovery time will be shorter and indirect losses will be less. The time of recovery also depends, in part, on how quickly and effectively recovery funds are spent.

It is only through mitigation that we reduce direct losses. It is largely through mitigation that we reduce indirect losses. The issue is not whether or not mitigation should be a mandatory element of a federal-private sector partnership in earthquake insurance; it is what mitigation can be used. There is no reason to wait. Mitigation should be an important insurance-industry strategy now. It is clear the insurance industry has to deal with its role in mitigation before we can talk about federal participation in earthquake insurance.

The insurance industry seems not to have the reserves available that they would like to have, and from my perspective, we do not have the insurance coverage we need. Because insurance is the traditional private-sector mechanism to spread the risk of financial losses without relying on government, it is a recovery mechanism that must be used more effectively. Resolving the problems that prevent its greater use is my interest.

In closing, my agency is writing a report on the Loma Prieta earthquake and will cover the financial impacts and the sources of funding. It should illustrate the cost of losses, and sources of recovery funds, for a pretty big earthquake, if not The Big One.

PRESENTATION OF ROBERT G. CHAPPELL

FEMA and California have been working pretty closely together on the Loma Prieta earthquake during the past year. Thank God Loma Prieta happened in California rather than South Carolina or in the central United States. California is the best-prepared state, both from a standpoint of state capability, local resources, search and rescue capability, and mitigation activities. It is a state that we all should look to for leadership and for guidance as the other states in the country that have similar earthquake hazards begin to develop the programs California now has. We see that as something that sets a good example for most of us.

With reference to Mr. Tobin's presentation, is that I am glad to hear California thinks this earthquake is going to cost less than we originally thought it was. We had just received $1.1 billion from Congress to deal with the Hurricane Hugo activities and then, 30 days to the day from Hurricane Hugo, Loma Prieta came along. Now, the question is: Would the Congress have readily made the second $1.1 million available to us on such a timely basis if Hugo had not occurred?

Certain folks wanted to fence in whatever funding was made available, either for Hugo or on Loma Prieta. The President's Disaster-Relief Fund, which we have the responsibility for managing, is a pot of money, it is no-year money, it is given to us on an annual basis by Congress, and we went in this case obviously for a supplemental appropriation, which can go to any disaster. Frankly, there were a lot of them this year in addition to Hugo and Loma Prieta. There was flooding throughout the Midwest. We had tornadoes. We are into hurricane season again, and heaven help us if we go through another Hugo at this point in time, because we are really still suffering from the administrative burdens of last year's activities.

I have a responsibility for managing the disaster program at FEMA. I am the civil servant guy. I have a boss named Grant Peterson who is the political appointee for the Disaster-Assistance Program, and we have been trying to do a kind of a dog and pony show over the past few months to deal with the lessons learned from Hugo and Loma Prieta and how we can all benefit from that.

It goes back many years, even into the 1950s, but it was 1973 where the Federal Disaster Assistance Relief Program was really created in what essentially is its present form. It came about as a result of a series of tornadoes that affected the central United States. One town in particular was the community of Xenia, Ohio, that was devastated by a tornado, and Congress at that point enacted a federal disaster-relief program, Public Law 93-288. This essentially established a federal relief program of substantial efforts to deal with immediate recovery activities and funding to the extent of dealing with temporary housing for disaster victims and individual and family grants in concert with the states to those that had unmet needs. The Small Business Administration had its own relief programs and in addition, FEMA's predecessor at that point was given the responsibility of working with state and local governments and restoring the disaster-damage infrastructure. That program was 75 percent federal, 25 percent state and local, although it originally started in 1973 and was 100 percent federal funding.

So in 1988, we saw a change in disaster-relief funding legislation, and the Robert T. Stafford Disaster-Relief Act was passed. Essentially, it made current what the 1973 law had brought us and established in law that it was not less than 75 percent federal funding for disaster relief to state and local governments, and then established a broad range of federal programs for FEMA dealing with individual and family needs as well as temporary housing, crisis counselling--many of the things that we have utilized during the past year in disaster relief activities.

The program is designed to deal with the immediate needs of families, individuals, and communities. It is not designed to deal with long-term needs. There has been very little effort and attention to the long-term recovery needs of the various communities, individuals, families, and industries that might be affected by disasters. The program that we have now on our plate, and which has been responsive to Hurricane Hugo, Loma Prieta, and some 30 other disasters during the last year, has really been a supplementary one. We are supplementary to the state and local efforts. We are not the first responder. State and local governments are the first responders. There are many in the federal community and in the Congress that believe that maybe the federal government should have some first-response capability--911, if you will, of the federal government--and FEMA has been asked to maybe take on that responsibility.

There are 233 people that work in the Disaster-Assistance Program, and who been spread during this year from the Virgin Islands and Puerto Rico to American Samoa, the Northern Marianas, California, Alaska, and various other places. Getting 233 people to deal with disasters in that broad range of geography is really not something that gives me any confidence we have the ability to be first responders. But simultaneously with all this disaster activity and with the changing of legislation, there was a series of events that were occurring outside this country. Two of those more significant ones were the Armenian earthquake that occurred a few years ago and the Mexico City earthquake, and it brought home the fact that really this country had a limited capability to respond to such catastrophic events.

These two instances brought life into an initiative to develop a program of federal response planning that dealt with a federal response to a catastrophic event. So in 1986, some 25 agencies and the American Red Cross agreed to develop a plan for federal response to a catastrophic earthquake, and this planning process has been going on during the past 3 or 4 years. Last August, almost a year ago, this plan was tested in California, and it was a Hayward fault earthquake exercise that was conducted in Sacramento, involving the state of California along with many federal agencies, local governments, and many other states being there as observers.

It was about an 8.0 earthquake that was premised along the Hayward fault which is on the east side of the San Francisco Bay. It was very successful earthquake exercise. One of the things it brought to us was a communication system allowing us and the state and local governments to know who the players were and how the process worked.

Well, it was good training because we not only had the Loma Prieta earthquake, but we had Hurricane Hugo. Many believed that Hurricane Hugo represented probably the closest thing to a catastrophic event that has been seen in this country in a long time. It was certainly catastrophic in the Virgin Islands, in St. Croix. We did not know how bad it was there, and because of poor communications we sent a team the day after the hurricane hit. When they arrived on site, they could hear gunfire in the distance, and there were no communications systems. They were to be our eyes and ears there, but little did we know that there were no support systems there, that local and state government--in this case one and the same--had broken down, and there were problems with law and order.

Simultaneously, the governor made a request for military troops to come in. The Department of Justice arranged for that, and at least we were able to deal with the safety of our people there, but there was a very severe breakdown of the capability of this particular governmental unit to provide services for its own people, as well as anybody else that might be affected.

We were in the process of providing immediate lifesaving equipment and supplies to the islands. We used the catastrophic planning process and the coordination process that had been developed with this Federal Response

Plan for a catastrophic earthquake as a base. There is a group called the Catastrophic Disaster Response Group, CDRG. The CDRG is the leadership of the various agencies that are involved, which are the signatory parties to the plan, and we got them all in our emergency information center on a 24-hour basis down at the FEMA building at 500 C Street. We then began to identify what were the needs of the people in the Virgin Islands, St. Croix in particular, and how could we deploy these resources. For example, we sent over 4 million pounds of food to the Virgin Islands. Some of those were meals-ready-to-eat, which I do not know if they ever were, because any of you who have eaten those types of things know that they are not too palatable, but if you are hungry, you will probably utilize them.

The concept was that our people would identify, working with state government, what are the needs, i.e., four-point-something million pounds of food. What types of food, etc. We would pass it off to the Food and Nutrition Service, which is a member of this CDRG. They would go and acquire the food. It would be delivered to a specific air station, and then the military would be tasked, all of this at FEMA cost, to fly it down to the Virgin Islands. Then it would be offloaded by the military that was on site, because they were the only resources available to us, and then distributed to Red Cross serving centers and other places there. Several generators, water purification equipment, and a complete mobile hospital unit (the hospital in St. Croix was damaged very badly by the disaster) were sent.

Portable toilets were also sent. That is one of the things that is not usually thought about a lot, but certainly it is a very basic need. Many thousands of pounds and many millions of dollars worth of equipment were deployed to St. Croix on probably a 3-week basis in order to stabilize the situation, and then state government was able to deal with the longer-term recovery.

The same thing, to a lesser scale, in Puerto Rico, and, of course, by that time, there were disasters in South Carolina, North Carolina, and, 30 days later, in California. So, to give you an idea of how that process worked, and the CDRG's responsibility, and the fact that the federal government does have an immediate-response capability that resides not only in FEMA with our 233 disaster employee resources but within the federal family. The plan that we have utilized is the Federal Response Plan. We also utilized portions of the plan in California when the Loma Prieta earthquake occurred, because we did not know the magnitude of it. We also did not know what the immediate needs were, so we convened the group and asked them to begin working with the state in identifying what the immediate needs were and whether or not California needed supplementary assistance that went beyond the norm, and frankly, they did not.

The Small Business Administration is our major ally in providing assistance to individuals and families. The purpose of their program is to provide assistance in the form of loans to not only businesses but to individuals and families that might be eligible for disaster-assistance loans. They are our major ally. They also spent many millions of dollars in the Virgin

Islands, Puerto Rico, and California, and they represent probably the long-term economic recovery effort that exists at this point in time as far as the federal government is concerned.

Now, we also have other programs at FEMA dealing with national security, but I cannot get into that. However, there is an element of capability in this country that has for a long time been concerned with the potential for a nuclear holocaust, and there has been a planning effort taking place as to how to continue government's capability in the event of such activities. And that involves long-term economic recovery as well as immediate needs being met. Until late last night, we believed that the world situation was cooling down sufficiently that maybe that was not going to be quite as important a fact, but as of this morning (Iraqi invasion of Kuwait), maybe we need to look at that again.

We also have been in contact with many members of private industry and business around the country, and they are doing their own catastrophic planning. AT&T is concerned about how to deal with catastrophic planning in the event of some natural calamity. We also have talked to the Disney people. We have talked to various other corporate organizations about catastrophic planning, how to go about it, the types of things they should be concerned about, and also try to be of assistance to their special needs.

Terrorism is something we do not like to think about, but certainly it represents the potential for a catastrophic event occurring in this country. We recently had a small exercise which dealt with that catastrophic event with the power grids being taken out around New York City, how is that dealt with, what are the implications of it, and certainly the social problems that might result from that and the economic and life-saving needs that would have to be dealt with. My observation of that exercise was that the plan that would be used to deal with such a thing once again resides in the Catastrophic Disaster Response Group that has dealt with Hugo, Loma Prieta, and probably would deal with such things as a terrorism incident in this country.

There has not been a great deal of planning, nor is there a full awareness of the implications of such events. Hopefully, it will not occur during my lifetime or my tenure in government, but I am concerned about it, and I believe that what I can see is that we are not prepared to deal with such events on an adequate basis, certainly on a long-term-recovery basis.

Approximately 40 requests a year come in for disaster assistance to the President. By the way, this is the President's disaster-relief program that we manage, and the President makes the decision as to whether or not a disaster is declared. Unfortunately, the director of FEMA gets to notify the governor when one is turned down. There are a number of disaster requests coming in from states that are not recognized by the President as meeting the needs of, and requirements of, the law.

Certainly, in the case of catastrophic or near-catastrophic events, the President always does declare these as major disasters. Last year, when Hugo was declared, we had a very limited amount of money in the bank. It happened in late September. We were awaiting the next appropriation, and we had inadequate funds to deal with the immediate needs of Hurricane Hugo victims, so Congress did give us a supplementary appropriation. Congress has been willing, interested, and very responsive to the immediate needs of disaster victims.

Will that continue? I am not certain. I am just not sure as to what the long range outlook is, particularly with the potential for sequestration of funds during 1991, with the Gramm-Rudman-Hollings provisions of law that the Congress is now dealing with in the budget talks, and what the long-term budget outlook is. If money is given to disaster assistance, it has to come out of somebody's pocket, come out of some other federal program, and this is one of the real difficulties that particularly Congress, certainly since Gramm-Rudman-Hollings, has had to deal with. Long range, it is not clear to me, but the experience would indicate that the Congress and the administration, too, have been very responsive to the disaster needs in this country.

A couple of other comments are that with the long-term nature of recovery, it is not easy in an earthquake to identify what the immediate needs might be and how to restructure your infrastructure. In a flood, you can go out and say, well, this bridge needs to be replaced and that road, and you can get in there and do it. In an earthquake, you cannot tell from the superficial nature of disaster damage whether the building is about to fall down or where there is just a hairline crack in that building. We are in very difficult, but I will not say adversarial, negotiations and discussions with, for example, Stanford University over some damage done to their buildings, as well as the city of Oakland concerning their city hall. The Oakland city hall is a historic structure. It was badly damaged. The city of Oakland would like to restore that structure to its predisaster condition and build it up to a certain code level, and that would require a great deal of federal investment. And our argument is that we either put it back to its predisaster condition or that we build a new facility, and the new facility would obviously be a lot cheaper than putting it back to its historic structure state, enhanced to current codes and ordinances.

So the big debate there is cost implications. We are talking about big dollars, and it is going to be a long-term process in making sure that these things are restored in a fashion where they do not represent a continuing hazard, and yet the federal investment is cost-beneficial and we meet the requirements of the law.

Litigation is also going on in California. Prior to the disaster occurring, there were folks there that were homeless, but the program deals with folks that were made homeless as a result of disasters. Yet it represents a national problem that somehow we are now involved in. We are involved in extensive litigation in California in providing housing for those that are in the fuzzy area of predisaster homeless, homeless during the disaster, and others that are now homeless since the disaster. This is another reason that it sometimes takes an inordinate amount of time to complete the recovery effort.

In conclusion, certainly mitigation is something that FEMA and the many representatives of our mitigation staff here today believe in very strongly. I came out of the National Flood-Insurance Program, as did Frank Riley, and Hal Duryee and we believe that the basic premise of flood insurance is that mitigation goes hand in hand with the provision of insurance. We are trying to promote that throughout the agency and certainly in the Disaster-Assistance Programs. Many requirements, provisions and funding for mitigation in the latest revision of thee Stafford Act. We will continue to try to promote that as a national public policy and we believe that, certainly in the state of California, we are interested in doing that as well.

The long-term economic recovery from catastrophic events is something that has not been dealt with adequately. FEMA has been concerned about this, but must first deal with the immediate recovery and response to the catastrophic event, and that is where our concentration of efforts has been. We do not have the resources to do much beyond that at this point in time, but the nation should consider what are our needs and what should be our future priorities.

PRESENTATION OF RICHARD J. ROTH, JR.

I am going to present two areas: the estimated insurance losses from the Loma Prieta earthquake and a short review of pending earthquake-insurance legislation in California.

Being a government regulator has some advantages. After the San Fernando earthquake in 1971, the Department of Insurance issued a special call for data to all of the licensed insurance companies in California. Since the Department of Insurance is a regulator, all of the insurance companies had to respond, so we got a 100 percent response. At that time, the companies reported a $46 million loss. That was such a good idea that when the Whittier earthquake came along on October 1, 1987 and was a 5.9 earthquake, that special call was repeated, and the reported insured losses were $73 million. This time, when the Loma Prieta earthquake came along, I issued a much more detailed special call, and I now have about five boxes of data. And it is not all in yet, because some of the companies have asked for extended time.

There is so much data that we have been going back to the companies and asking them to put it on computer diskettes so we can analyze it. However, for this presentation I took the largest 44 companies that we had and came up with the following summary. (The attached summary is a complete compilation of all companies from a subsequent second call and is the final compilation. The final results show that the total losses to the insurance industry will be over $901 million. The compilation of 44 companies is not attached since it is not obsolete.) For the Loma Prieta earthquake, if you make an allowance for the fact of the deductible plus you take into account that only about 25 or 30 percent of the people were insured, you can scale up to these figures to get an estimate of the total damage caused by the earthquake, insured and uninsured. This would be about $5 billion or $6 billion. In the case of the Whittier earthquake, less than half of the losses were excluded by the 10 percent deductible and 25 percent of the homes and commercial structures were insured. Therefore, if the insurers paid out $75 million in insured losses after the deductibles, then the total losses, insured and uninsured, would be approximately %75 x 1.67 x 4 = $500 million dollars.

One interesting fact is that the Whittier earthquake was roughly a size 6 earthquake. The Loma Prieta was roughly a size 7, and so there seems to be a scaling of 10 as you go up on the Richter scale. Even though the Richter scale increases about 30 for energy release, it seems that there is a size 10 scaling. The Whittier earthquake is roughly $500 million, the Loma Prieta $5 billion, and the Earthquake Project is projecting $50 billion for a size 8. It seems that we are getting a scaling factor of 10, and this data confirms that.

Concerning the Loma Prieta data. This was also the first time ever that any regulator asked life and health insurance companies for data on life and health insurance, and we received 5 death claims for a total incurred loss of $498,000, 21 accident and health claims for a total loss of $828,000. Now, we do not know whether we got all the A&H claims, because the life insurers are not set up to handle that, but they will be trained. The questionnaire asked not only for the number of claims with payments but also the number of claims, whether or not there was a payment; and there were 97,000 claims filed with insurers, of which 47,000 involved a payment. The reason there were nonpayments is because they were below the deductibles, and the deductible, in most cases, was 10 percent. What was amazing was that there were 97,000 claims and only 47,000 required payment.

For earthquake coverage losses, there were 24,000 claims with payment, and the total amount of payment was $320 million. There were 6,001 automobile claims filed upon which there was a payment--7,500 claims total for automobile, but there were 6,001 cars insured for which they filed a claim and there was a payment--and the total payments were $7,900,000, so that is over $1,000 average per car. This gives an estimate of the vast number of claims there are, even in an earthquake that does not seem to be quite as devastating in terms of the number of claims as the Loma Prieta was.

In terms of fire losses, 201 claims were filed and 183 claims were paid. Over $10 million was paid on fire losses. Over $2 million on just homeowner's alone. I am going to analyze all of this data and issue a special report sometime in the future. This data includes not only the figures given here, but also a listing of all of the claims by ZIP code, so I will know for each ZIP code the damageability factor. I also will know by ZIP code the number of policies the insurance company issued and the number of policies upon which there was earthquake coverage. This has told me that in the San Francisco Bay area, 30 to 40 percent of the homeowners had earthquake coverage. This is larger than the state average. The state average is about 25 percent.

Also, I asked for the detailed information for homes so I will have detailed information on structures, contents, and what is called temporary living expense. For business I will have detailed information on structures, inventory, and business interruption. We know very little about commercial buildings, particularly small businesses, and we know almost nothing about the business-interruption coverage. On homeowners, the temporary living expense is a very important coverage, because that kicks in when the person has to leave the dwelling or cannot occupy the dwelling, either because it is damaged or because there is a police line around it. From my point of view, that coverage has virtually no deductible, whereas the regular insurance has a 10 percent deductible.

Let me see if I can give you some quick differences between a homeowner's coverage and a small business. Homeowner's, as mentioned, make up about 30 to 40 percent of the homes insured. The average earthquake coverage was between $100,000 and $125,000. About 6 percent of those insured with earthquake coverage had paid loss. The average loss after the deductible was between $10,000 and $15,000, and the total loss amounted to less than 1 percent of the value of the insured homes. For small businesses, you get a little different profile. For small businesses, less than 10 percent of the small businesses had earthquake coverage, so there are substantially fewer small businesses with earthquake coverage. In fact, one of the major agents in Watsonville said that he did not know of anybody who had earthquake coverage on small businesses.

Most of the small businesses that were damaged that he was in contact with had no coverage whatsoever. On a small business, those that had coverage, the coverage was about twice what it is for a home; in other words, about $200,000 for earthquake coverage. And also, the average loss was about twice as great for a home. It was about $20,000 to $30,000 for a small business. This is the only data I am aware of that exists on small businesses. I also have all businesses, but basically my primary interest is small businesses.

We also did a survey of insurance companies and asked them the number of policies statewide with earthquake coverage, and we came up with a result which interested me. Statewide--we are getting off Loma Prieta for a minute--statewide, homeowner's policies were 23 percent. Condominiums, 26 percent of the condominium owners had earthquake insurance and renters had 26 percent. What interested me is that I had no idea so many condo owners and renters had earthquake coverage. What this tells me is that the demand for earthquake insurance is broadbased. It is not just homeowners protecting equity, it is a broader social demand for earthquake insurance. Risa Palm is doing a study on the demand for earthquake insurance. She also has concluded that the demand for earthquake insurance is broad based over all age groups, educational groups, income levels, and the amount of equity you have in your home. It really is not correlated with any of those factors. It is simply based on a demand, an individual demand for earthquake coverage, and that demand is broad based, as shown by these figures and also by the fact that in the San Francisco Bay area, as I mentioned, over 30 percent of the people have earthquake insurance, and they are paying quite a bit, $200 to $400 for this coverage.

Now, there is an economic question that occurred to me in these figures. I was interested in the fact that the percentages here have been practically the same for homeowners, condos, and renters; in other words, about 25 percent.

The premium, the average premium paid, is $217 for homeowners, $69 for condos, and $39 for renters. Another factor may be entering in here, and that is the price that was charged by the insurance company. When they go to set their rate, I have a feeling, although I cannot prove it, that they set their rate to try and get a balance, so that they only get about 25 percent of their insurers. I detect in working with the insurance companies that when they start getting a penetration greater than 30 percent, what they do is they start tightening up on their underwriting, they start raising their rates, and very few insurance companies like to have a book of business greater than that. What they are looking for is a spread, and they do not want a high concentration of insureds buying earthquake insurance. You see this particularly in San Francisco, where we get a lot of complaints in the department, where somebody has a house and they were turned down or nonrenewed for their earthquake coverage, and the reason is simply the insurance companies are limiting their exposure in that area.

Going on to another topic, I work closely with Karl Steinbrugge, who also works closely with the U.S. Geological Survey. We recently published a book entitled *Earthquake Losses to Single Family Dwellings, California Experience*, and I want to express my gratitude to the U.S. Geological Survey for printing this. This book's contents are available from the U.S. Geological Survey. The book lists extensively the data and the loss curves, loss-over-deductible curves, for the San Fernando and Whittier earthquakes, and this is particularly important for insurance purposes, because it gives the amount of loss that the insurance company can expect for a 5 percent deductible or a 10 percent deductible or a 15 percent deductible and also for a zero deductible, so it gives the total loss that is expected for a particular type of building, a particular type of home.

The expected loss varies dramatically whether it is pre-World War II or post-World War II or whether it is wood frame or masonry or it is built on wood foundation or a concrete foundation. This book contains all the data that I am aware of. This is raw data--this is not theoretical--this is raw data on dwellings, and also it contains attenuation curves from distance to a fault, and it also contains formulas for scaling up the losses for larger earthquakes. We therefore feel that in California, we know a lot about dwellings and the impact of an earthquake on dwellings. We do not know very much about the impact of an earthquake on small businesses or large businesses, and that is an area where we need to do more research. Unfortunately, it is a dramatically more complicated subject.

I also issue an annual report. I send out an additional questionnaire to all licensed companies automatically annually, asking for their exposure and the amount of business they have written by construction class. Using that information, I write an annual report every year. What this report does is give losses to structures. It does not apply to workers' compensation, automobile, or life and health, just structures. What it does is it gives the industry's exposure to structures, but it also attempts to look at the economic impact.

For instance, in Los Angeles, my latest report shows that the insured losses are about $6 billion. This breaks out to commercial, $5 billion, and residential, $1 billion, so even though the residential gets a lot of attention, actually the main exposure in terms of dollars is the commercial. Now, in insurance we have what is called reinsurance, so that when an insurance company insures a large building, they do not keep the whole risk; they turn around and reinsure it with other companies. Over 60 percent of the commercial business is reinsured around the world, and less than 23 percent of the residential. There is much less need for residential insurance to be reinsured, because you have a smaller risk, and more of them, whereas in commercial you have a higher dollar value, and there are fewer of them, so you need to spread the risk among insurance companies.

Now, one of the main reasons for the Earthquake Project is the industry is running out of reinsurance capacity. We are dealing with a worldwide community. If there is an earthquake in San Francisco, it is not just California companies that will pay for it, and not even United States companies that will pay for it; it is the whole world financial insurance market that will pay for it, and we have just about reached the point where there is just no more reinsurance.

One of the legislators asked me, "Well, can't we create a fund and use that fund to buy reinsurance in the world market?" I made a few telephone calls, and all of them gave me the same answer. They said the most the state could buy would be $250 million of coverage for reinsurance if they set up a program. That gives you an idea of how limited the reinsurance capacity is. I know the people in New Zealand were very concerned because of their exposure. For a while, it was just limited to the capacity of the government and a fund they built up to pay for an earthquake loss in New Zealand. They said they felt that what they should do is buy insurance in the world market. So they went out and with a great deal of difficulty, they bought $1 billion worth of coverage for risks in New Zealand. Reportedly, they have just bought with great difficulty that second billion dollars of coverage in the world market, and so the insurance industry just cannot really expand its coverage of commercial, because there is not any reinsurance market available.

Another point I want to make is that I just mentioned that the insurance industry insures $1 billion of residential coverage. Okay, let us do some simple mathematics. That residential coverage has about a 10 percent deductible. If you use the damage curves that are in Carl's book, going from a 10 percent down to zero will increase the losses about five times. In other words, if you were to insure everything for a zero deductible instead of a 10 percent deductible, you would have five times more loss. So we multiply that times five. Now, if only 25 percent of the people have insurance, to get the total loss, insured and uninsured, you have got to multiple by four. So if you multiply $1 billion times five times four you get $20 billion. That is the total estimated loss to dwellings from one earthquake, one size 8 earthquake in California, so that is a way of estimating what the total loss would be just to dwellings, and that is not counting all the other coverages, workers' compensation, automobile, and commercial.

What happened after both the Whittier and the Loma Prieta earthquakes was that the legislators received many, many complaints about the 10 percent deductible. I do not think there is anything that bothers people in California more than the 10 percent deductible. Yet, they understand it. You talk to somebody, they say, yes, we understand it. The industry does not want to settle all these small losses. We just would not have coverage unless there was some kind of large deductible like that. It is basically a catastrophe coverage, but even though they say that, the reality is that when they have about $200,000, of coverage so that means a $20,000 deductible, if you have a loss and you have to pay the first $20,000 that is a significant hardship for a lot of people.

So there were a lot of complaints about the 10 percent deductible. Another thing that happened is that the legislators took notice of the fact that there were a lot of disaster-relief loans and grants to be paid out. After the Whittier earthquake, the loans and grants amounted to at least $175 million. The Loma Prieta earthquake--as Tom Tobin says, he does not know, and I do not know--but it is at least a billion dollars. What happened is that after the major earthquake, there is a huge demand on the state legislature and on the federal government for disaster-relief loans, so you have this sudden, unexpected, unwelcome demand on the financial resources of the state.

A third issue that came up was that the Northern Auto Club had a $1,500 coverage for temporary living expense. It turned out that as small as this is, only $1,500, it was immensely popular, and the policyholders of the Northern Auto Club were ecstatic to get this $1,500 check. And the legislators took notice of this and saw that for very little money, you can make a lot of people happy.

A fourth issue that came up was the raising of the sales tax 1/4 percent just to pay for the road damage. They could not pay it out of their own resources. They had to raise the sales tax, and we know what the perils are of asking for a tax increase, but the people accepted it. Anyway, the point is the state resources were so limited that they had to raise the sales tax. Putting all of these together, California legislators say we have got to do something, we have got to prepare for the future. We have to have a prefunded insurance program in advance of the next earthquake of the size of Loma Prieta. So the legislators fell all over themselves submitting bills. I am not kidding, the pile of bills that high covering everything.

There are four main bills dealing with insurance, and I am not going to go into detail, because they have a lot of common features. The common features are all prefunded and what they want to do is collect the money, put it into the state fund, and have it accumulate taxfree. Now, they know that this is legal because the workers' compensation state fund does not pay federal taxes, so there is an insurance program which operates free of federal taxes, and the California workers' compensation fund is a residual market. If you cannot get workers' compensation anywhere, you can get it from the state fund, and that is the reason for its existence. So you can create a fund and have it accumulate taxfree.

Now the question is, what kind of coverage do you want? The governor made a proposal to have $15,000 of coverage, $15,000 excess of the deductible, roughly of $1,500. There is a formula but it is roughly $1,500. The idea is that, the current state of the proposal is that everybody with a homeowner's policy would pay a surcharge--it is not a tax, not a premium, it is a surcharge--and if you call it a premium then you have to pay a premium tax on it. If you call it a tax, that is bad. It is a surcharge, and it is going to be put into this fund and then accumulate. Now, they want this fund to build up, and they want to offer this coverage, but they do not want the state to be liable. So they have what is called the Japanese formula in there in that if there is a loss, the fund essentially goes bankrupt and pays a pro rata share. So if the fund is a billion dollars and the total losses under this program are 2 billion, then you get 50 cents on the dollar.

The reason for the $15,000 coverage is they wanted a fairly small amount, but they did not want to compete with the voluntary market, so that if you want full coverage, you would still have to go into the voluntary market. And also, they did not want to conflict with any federal program which might be set up. In fact, they explicitly support the federal program. There is a Senate Joint Resolution 57 in the Legislature which supports any federal earthquake insurance program. They urge Congress to pass that.

Other features of these four bills, and they vary somewhat, are the use of money for retrofitting and mitigation. Most of the bills have quite a bit to do with retrofitting and mitigation. The governor's bill has a provision in there where if the amount starts to exceed a billion dollars, then a certain percentage of the fund can then be used as loans for retrofitting and so forth, which is an excellent idea, and as the fund grows larger and larger, a larger and larger proportion of that fund is legally available for retrofitting and mitigation.

GENERAL DISCUSSION OF CHAPTER 7

QUESTION: I have a question for Dick Roth. Dick, it is a real pleasure to work with you and have someone as interested as you are in the insurance area. When you talk about the loss to dwellings, I think you projected a figure of $20 billion. Is that a vibration damage only or does that include fire following and additional living expenses?

MR. ROTH: No, that is strictly structure, not fire following. It was based strictly on the structural losses estimated in my questionnaire, expanded to remove the deductible and expanded to include everybody who is insured and noninsured. Fire losses would be in addition.

QUESTION: Additional living expense also would be in addition to the coverage.

MR. ROTH: The additional living expense should be extra, but I have a feeling some companies are including potential liability for additional living expense in the estimates they give to me. But the additional living expense should be additional.

QUESTION: I have a question for Mr. Tobin. What types of losses do you expect to residential dwellings in California?

MR. TOBIN: In California, the primary type of dwelling is a wood-frame dwelling. The primary weakness we have in about 25 percent of our 6 million or so dwellings is they are not bolted to their foundations. They were constructed prior to the time the code required that. Evidence is that will cost from 1 to 2 percent of the building's value to bolt them to the foundation. After the Loma Prieta earthquake, it cost 10 to 15 percent to put them back on their foundations. If you have a weak soft story like in the Marina District, or cripple walls, the cost of retrofitting would be higher. Usually you have fairly good access, so it is not too expensive. When you deal with brick buildings, multifamily residences, the costs go up. The cost to retrofit unreinforced masonry buildings is 10 to 25 percent of the replacement value of the building.

QUESTION: This creates a problem then. If we give them a discount in premium for retrofitting, premiums will have to be raised for those not retrofitted.

MR. TOBIN: That is true if premiums are risk based. However, the requirements for mitigation need not be solely an insurance premium. For example, there is a proposal by a state senator to require insurance companies to provide a homeowner's booklet explaining how to identify weaknesses in their houses and ways to strengthen them. The insurance industry is neutral on that bill. That is unbelievable! It's such an easy way for the insurance industry to support a mitigation program.

QUESTION: Mr. Tobin, you said it costs 10 to 15 percent of the value of the house to put it back on the foundation?

MR. TOBIN: Yes.

QUESTION: And that is about what the deductible was?

MR. TOBIN: That is correct. There is an interesting study relevant to your question. Ron Gallagher, who is a structural engineer, did a survey after the Loma Prieta earthquake of the costs of repairs to single-family houses and mobile homes. That study developed some cost-benefit ratios. It is interesting data, and I recommend it to you. It can be obtained from Mr. Roth's agency.

QUESTION: I wanted to ask Mr. Roth, from the legislators' point of view, how do they feel about two issues: one, what is the government's view of universal mandation? and number two, how do they feel about having a surcharge that is fully risk based?

MR. ROTH: Most of these bills are mandated in the Legislature. In other words, the surcharge is put on every homeowner's policy. About riskbased, the provision in the governor's bill is that it would be risk based according to soil condition, age, the characteristics of the house, but not the value. Not the value because it is a limited, $15,000 policy. The intent is that it should be actuarial.

The governor's bill happens to have a range or a limit. In other words, it must be greater than this and less than that; but that is subject to change over time. Some of the other bills do not have a range in them. So actuarially, the range is dramatic. It can go from practically pennies for a well-built house up in the northern counties to hundreds of dollars for a beautiful brick home in Berkeley.

QUESTION: I would like to direct my question to Mr. Roth. Could you please elaborate on that kind of limited insurance that is being proposed. Is it basically trying to take care of the deductibles that are presently in the insurance policies? It seems to me that if that particular fund is actuarially fair and if it will be able to pay the insured claims, that the smaller claims are more easy to predict in terms of recurrence and in terms of the capability of the insurance industry to honor their policy. Is it really just a matter of tax purposes? It seems to me that it should be the other way around. If the insurance industry cannot handle big claims, which are the ones that they are now having, then there is a reason to call in the federal government. But when you are talking about small claims, 20 percent deductible, 10 percent deductible, and the industry's ability to deal with them, I do not think that is the right way to approach it.

MR. ROTH: You have to step back and ask: what are the options? What the legislators saw was an opportunity to have the maximum political benefit for the buck. Yes, it was designed to cover most of the deductibles, and this created a lot of the problem. Also, they wanted to give us, because of the popularity of the Northern Auto Club program, at least some money. It does create a problem, because you are dealing with a segment where the frequency is dramatic. As I mentioned, there were 100,000 claims filed for all different coverages. I consider that a lot of claims. If you are insuring the $15,000 that way, you are going to have a lot of claims. That is one of the reasons why the coverage in the bill was limited to structures only. Currently, insurance covers structures and contents.

MR. TOBIN: I think there is another issue, and that is that we really do not know where to invest the money to speed recovery. It might be a wiser recovery strategy to insure small businesses rather than invest small sums of money into damaged residences. We just do not understand the recovery process well enough.

There is another issue that came up as the insurance bills were being heard. Another bill would have provided coverage up to $100,000, also for a low premium. It had a similar formula: It would have been mandated, and if the amount in the fund was less than the total claims, there would have been a proration. The reaction of the chairman of the insurance committee was that if the fund were not fully funded and homeowners had been encouraged to back out of the commercial market for catastrophic coverage, they would be shocked to find out that they might only get $5,000. He said that this situation would be "tantamount to a fraud on the public," whereas if the

$15,000 coverage is not fully funded and the payment is prorated to only $5,000, he said "that is just a bad day at the office." Politically, it was a lot easier to go the route of a "deductible gap" insurance.

QUESTION: To return to the homeowner's, the bolted foundation and unbraced cripple wall really represents a tremendous problem, because that is your $15,000 loss. But it also means that the home is unusable for weeks or months, which is a tremendous social hardship. There are people in Watsonville today still living in shelters because of that situation. At the same time, it is the one mitigation effort with the clearest benefit. You can almost ensure that will not happen if you put plywood on the cripple walls. It is also very easy to do yourself. If you do it yourself, the cost is about 1 percent. I would like to see a little more imagination applied to getting that done. In other words, what I have in mind is something which relatively unskilled people could do. You could have the California Conservation Corps trained to do it. In the energy programs, there is a lot of imagination used in terms of grants and loans to get people to insulate their dwellings. So let us look at changing that part of the environment as well; and, if we find something is worth doing, let us also find imaginative ways to do it. I think that is a very big payoff on its own.

MR. TOBIN: There is a bill on the governor's desk today that would require that homes be bolted or that crippled walls be strengthened on sale or transfer. It would be a cost that the seller would have to bear. It is another bill where the insurance industry is neutral, yet we need a governor's signature on that bill. There is a tremendous potential savings there for the industry. [Note: The bill was vetoed.]

QUESTION: Is the deductibility of casualty losses and the effect of tax payments making people who claim such losses consider it income provision?

MR. ROTH: No, I do not look at that at all.

QUESTION: But yet presumably all these losses that are not covered by insurance are tax deductible. You might want to comment on that. That, after all, represents a transfer from general to federal revenues to taxpayers in California who experienced that loss.

MR. ROTH: Yes, the federal government is in the business of insuring for earthquakes right now. For that reason, there are disaster-relief programs.

MR. CHAPPELL: That is correct. I think that not only the benefits that might accrue, if you want to call it benefits, from the income tax provisions, but also the unmet needs otherwise not covered by insurance or other resources that might be eligible for some type of federal assistance, also represent costs to the taxpayer.

References

1. Federal Emergency Management Agency (FEMA). 1990. Estimated Future Earthquake Losses for St. Louis City and County, Missouri. Earthquake Hazard Reduction Series 53. Washington, D.C.

2. FEMA. 1985. An Assessment of Damage and Casualties in the Central United States Resulting from Earthquakes in the New Madrid Seismic Zone. Washington, D.C.

3. USGS. 1983. Proceedings of a Workshop on the 1886 Charleston, South Carolina, Earthquake and its Implications for Today. Open File Report 83-843. U.S. Geological Survey. Reston, Virginia.

4. Metropolitan Boston Area Earthquake Loss Study: Loss Analysis. 1990. Committee Report to the Massachusetts Civil Defense Agency. Boston, Massachusetts.

5. D.G. Friedman. 1969. Computer simulation of the earthquake hazard. Pp. 153-181 Geologic Hazards and Public Problems Conference Proceedings. Office of Emergency Preparedness, Executive Office of the President. Washington, D.C.: U.S. Government Printing Office.

6. M. Kaplan. 1972. Actuarial aspects of flood and earthquake insurance in The Proceedings of the Conference of Actuaries in Public Practice, XXI. Chicago, Illinois.

7. D.G. Friedman. 1984. Natural-hazard risk assessment for an insurance program. The Geneva Papers on Risk and Insurance, International Association for the Study of Insurance Economics (Geneva Association). Geneva, Switzerland. 1(30[Jan.]):57-128.

8. D.G. Friedman. 1988. U.S. Hurricanes and Windstorms. London: Insurance and Reinsurance Group, Ltd. 270 pp.

9. D.G. Friedman. 1990. Is Hugo a forerunner of future great hurricanes? Research Review, Journal of the Society of Insurance Research. 3(1[July], 2[Oct.]). Appleton, Wisconsin.

10. USGS. 1985. Estimation of Earthquake Effects Associated with Large Earthquakes in the New Madrid Seismic Zone. Open File Report 85-457. U.S. Geological Survey.

11. S.T. Algermissen. 1983. An Introduction to Seismicity of the United States. Earthquake Engineering Research Institute. Berkeley, California.

12. R. Street and A. LaCroix. 1979. An Empirical study of New England seismicity: 1727-1977. Bulletin of the Seismological Society of America. 69(1):159-175.

13. AIRAC. 1989. Surviving the Storm: Building Codes, Compliance, and the Mitigation of Hurricane Damage. Insurance Research Council. Chicago, Ill.

14. USGS. 1990. Tecumseh's Prophecy: Preparing for the Next Great Earthquake. United States Geological Survey Circular 1066.

15. AIRAC. 1987. Fire Following Earthquake--Estimates of the Conflagration Risk to Insured Properties in Greater Los Angeles and San Francisco by C. Scawthorn. Insurance Research Council. Chicago, Ill.

16. S.P. Nishenko and G.A. Bollinger. 1990. Forecasting Damaging Earthquakes in the central and eastern United States. Science (September 21):1412-1415.

17. AIRAC. 1988. Earthquake Losses Under Workers' Compensation and General Liability--Estimates for a "Worst-Case" Event in Greater Los Angeles, by D. G. Friedman. Insurance Research Council. Chicago, Illinois.

18. NRC, Panel on Earthquake Loss-Estimation Methodology. 1989. Estimating Losses from Future Earthquakes. Washington, D.C.: National Academy Press.

19. W.J. Petak and A.A. Atkisson. 1982. Natural Hazard Risk Assessment and Public Policy: Anticipating the Unexpected. New York: Springer-Verlag.

20. K.V. Steinbrugge. 1982. Earthquakes, Volcanoes, and Tsunamis: An Anatomy of Hazards. New York: Skandia America Group.

21. R. Gulliver. 1986. Estimation of Homeless Caseload for Disaster Assistance Due to an Earthquake. Washington, D.C.: Federal Emergency Management Agency.

22. W.E. Spangle, R.L. Meehan, H.J. Degenkolb, and M.L. Blair. 1987. Pre-Earthquake Planning for Post-Earthquake Rebuilding. Southern California Earthquake Preparedness Project. Pasadena, California.

23. Applied Technology Council. 1985. Earthquake Damage Evaluation Data for California. Redwood City, California.

24. R.T. Eguchi, K.J. Tierney, and R.A. Antonoplis. 1988. An assessment of the risk of hazardous materials release during earthquake--the impact on surrounding communities. Paper presented at the Ninth World Conference on Earthquake Engineering. August 8. Tokyo.

25. K.J. Tierney, H.A. Seligson, and R.T. Eguchi. 1990. Chemical Hazards, Mitigation, and Preparedness in Areas of High Seismic Risk: A Methodology for Estimating the Risk of Post-Earthquake Hazardous Materials Releases. Draft report prepared for the National Science Foundation and the National Center for Earthquake Engineering Research.

26. C. Perrow. 1984. Normal Accidents: Living with High-Risk Technologies. New York: Basic Books.

27. A. Mazur. 1973. Disputes among experts. Minerva 11:243-62.

28. P. Slovic, B. Fischhoff, and S. Lichtenstein. 1981. Perceived risk: psychological factors and social implications. Pp. 17-34 in Proceedings of the Royal Society of London, A 376.

29. H.C. Cochrane. 1975. Natural Hazards and Their Distributive Effects: A Research Assessment. Boulder: University of Colorado, Program on Technology, Environment, and Man.

30. C. Scawthorn. 1987. Fire Following Earthquake: Estimates of the Conflagration Risk to Insured Property in Greater Los Angeles and San Francisco. All-Industry Research Advisory Council. Oak Brook, Illinois.

31. J.P. Brown. 1972. The Economic Effects of Floods. New York: Springer-Verlag.

32. J.D. Schilling, C.F. Sirmans, and J.D. Benjamin. 1989. Flood insurance, wealth distribution, and urban property values. Journal of Urban Economics January:43-53. 26(1):43-53.

33. D.S. Brookshire, M.A. Thayer, J. Tschirhart, and W.D. Schulze. 1985. A Test of the Expected-utility model: evidence from earthquake risks. Journal of Political Economy April. 93(2):369-389.

34. R. Palm. 1990. Natural Hazards: An Integrative Framework for Research and Planning. The Johns Hopkins University Press. Baltimore, Maryland.

35. R.L. Bernknopf, D.S. Brookshire, and M.A. Thayer. 1990. Earthquake and volcano Alerts: an economic Evaluation of risk-perception changes. Journal of Environmental Economics and Management. 18(1):35-49.

36. P.J. May. 1985. Problems in Formulating Disaster Relief After Mount St. Helens, Final Report to the National Science Foundation. Seattle: University of Washington.

37. J.D. Wright and P.H. Rossi, eds. 1981. Social Science and Natural Hazards. Cambridge, Massachusetts: Abt Books.

38. H.C. Cochrane. 1975. Natural Hazards and Their Distributive Effects. Monograph #3. University of Colorado, Institute of Behavioral Sciences.

39. K. Erikson. 1976. Everything in its Path: The Destruction of Community in Buffalo Creek. New York: Simon and Schuster.

40. H.P. Friesma, J. Caporaso, G. Goldstein, and R. Lineberry. 1979. Aftermath--Communities After Natural Disasters. Beverly Hills, California, and London: Sage Publications.

41. E. Haas, R.W. Kates, and M.J. Bowden. 1977. Reconstruction Following Disaster. Cambridge: Massachusetts Institute of Technology Press.

42. A.H. Barton. 1969. Communities in Disaster: A Sociological Analysis of Collective Stress Situations. New York: Anchor Books.

43. D.C. Dacey and H. Kunreuther. 1969. The Economics of Natural Disasters: Implications for Federal Policy. New York: Free Press.

44. J.D. Wright, P.H. Rossi, S.R. Wright, and E. Weber-Burdin. 1979. After the Clean-Up: Long-Range Effects of Natural Disasters. London: Sage Publications.

45. P.H. Rossi, J.D. Wright, and E. Weber-Burdin. 1982. Natural Hazards and Public Choice. New York: Academic Press.

46. J.V. Henderson. 1977. Economic Theory and the Cities. New York: Academic Press.

47. W. Stull. 1974. Land use and zoning in an urban economy. American Economic Review. June:64(3):337-347.

48. A.M.J. Yezer and C.B. Rubin. 1987. The Local Economic Effects of Natural Disasters. Paper #61. University of Colorado, Institute of Behavioral Sciences.

49. R.W. Ellson, J.W. Milliman, and R.B. Roberts. 1984. Measuring the Regional Economic Effects of Earthquakes and Earthquake Predictions. Journal of Regional Science. 24:559-579.

50. L.J. Stern. 1990. Insurance Stock Trends. Best's Review, life/health insurance edition. 90(10[Feb.]).

51. P.E. Wish. 1990. Review and Preview: 1989 and 1990. Best's Review, property/casualty insurance edition. 89(9[Jan.]).

52. Tokai Bank. Southern Kanto Big Earthquake and Japanese Economy. R&D Department, excerpt translation. Undated.

53. The Earthquake Project. 1989. Catastrophic Earthquakes: The Need to Insure Against Economic Disaster. National Committee on Property Insurance.